BACKYARD GUIDE TO THE NIGHT SKY

SECOND EDITION

BACKYARD GUIDE TO THE NIGHT SKY

SECOND EDITION

ANDREW FAZEKAS

NATIONAL GEOGRAPHIC
WASHINGTON, D.C.

CONTENTS

Veil Nebula, the remnant of a supernova

PREVIOUS PAGES: The Milky Way arcs over Organ Pipe Cactus National Monument in southern Arizona.

ABOUT THIS BOOK

EACH OF THE 10 CHAPTERS found in this new edition of *Backyard Guide to the Night Sky* is overflowing with information to guide you into the world of astronomy. Starting with the basics of stargazing, the book carries you from the inner solar system, on to neighboring planets, and into the deep sky with its black holes and supernovae. Along the way, it helps you learn to find and identify the 58 most important constellations visible from the Northern Hemisphere, both the stars that create these classic shapes and also the galaxies and other deep-sky objects within them, bringing a new sense of wonder to a night spent outside.

Not only will you find practical advice on how to recognize constellations and planets in the sky, but also you'll learn about discoveries from recent space missions and current scientific theories about celestial objects, how they were formed and what they represent in the grand scheme of the cosmos.

Each page includes fascinating facts and figures to enhance your appreciation of the night sky, such as the history of astronomy, the mythology of the constellations, and the key accomplishments of sky-watchers through the ages. Look to the featured essays inserted between the chapters for practical tips and advice, such as how to shop for binoculars or photograph glittering spectacles in the night sky.

Further Facts

These boxes contain fascinating facts, short histories, and cool things to spot in the sky.

Sidebars

There are three different types of sidebar. "Sky-Watchers" sidebars tell the stories of key astronomers and scientists, "The Story of" sidebars present legends and lore, and "The Science of" sidebars explain the science behind night sky observations and phenomena.

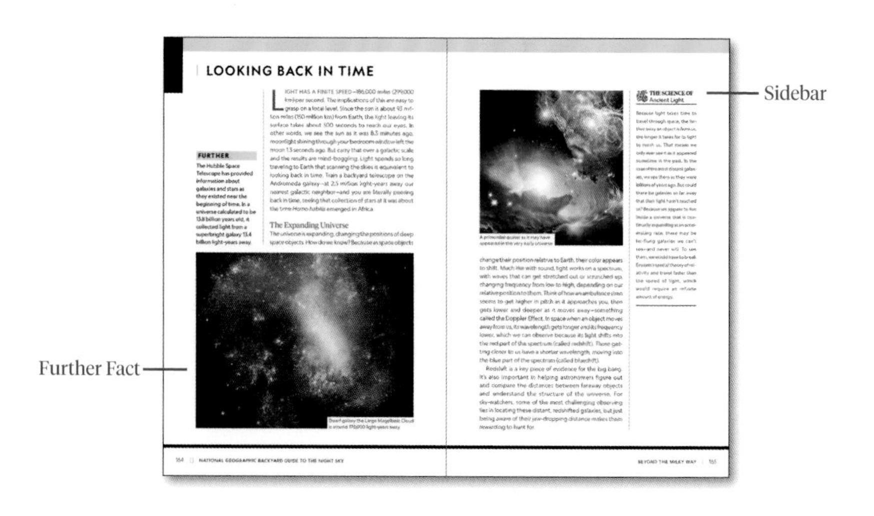

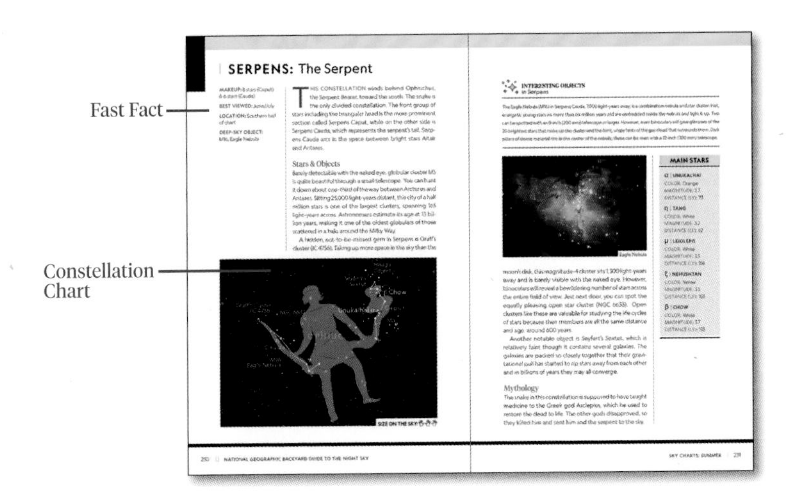

Fast Facts

These lists provide a quick reference to the key facts and figures about night sky objects. Tables will also list fast facts about sky objects to look for, such as annual meteor showers or visible star clusters.

Constellation Charts

The final chapter of this guide presents a grand tour of 58 constellations, each depicted with its own chart, created exclusively for this book, with artwork interpreting the mythological figure it represents. Each chart features the shape of the constellation, its main stars visible to the naked eye, deep-sky objects to observe through binoculars or a telescope, and neighboring constellations.

Seasonal Sky Charts

Each season brings a different piece of the sky into view. Sky charts for spring, summer, autumn, and winter show the full scope of what is visible to a stargazer in North America. Following each season's sky chart you will find a companion star-hopping chart, with directional arrows that illustrate a tactic for using familiar stars and shapes as signposts for finding fainter constellations and deep-sky objects.

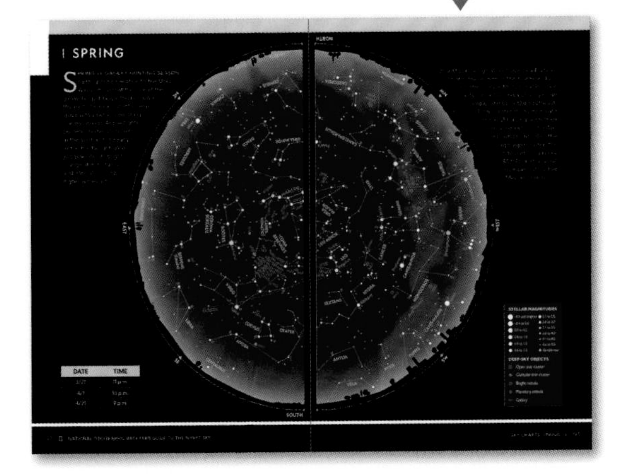

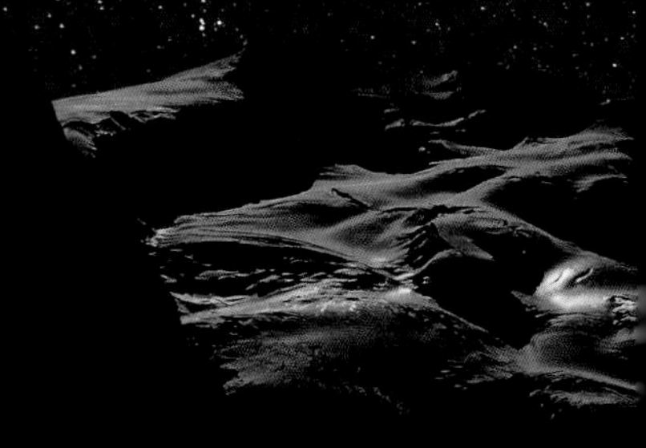

A field of stars shines brightly in a valley far from electric light.

CHAPTER 1

DISCOVERING STARRY SKIES

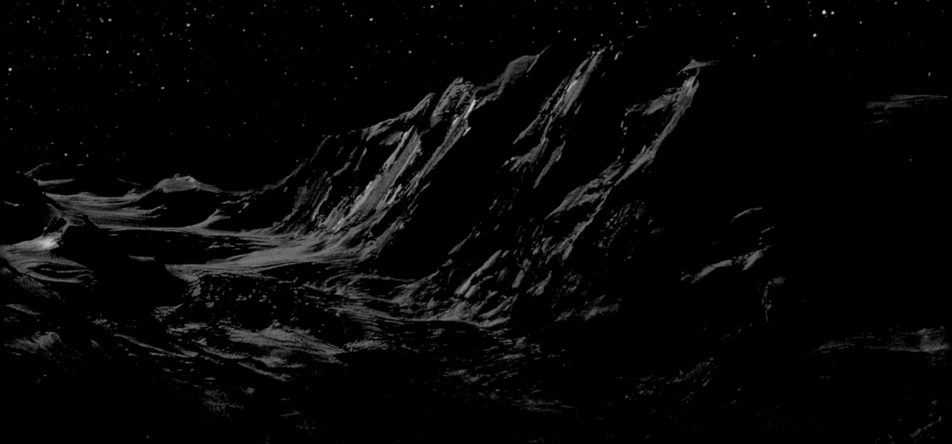

INTO THE DARK

A new star shrouded in gas and dust

WITH ITS GLITTERING BEAUTY, the night sky is considered one of our longest-standing sources of entertainment, news, and wonder. The parade of constellations, the undulating glow of an aurora, and the blaze of a comet have fascinated humankind for as long as we've been watching. The stars gave early civilizations information about issues as critical as when to plant crops and harvest them, and they helped them navigate the world.

Sky-watching remains an accessible hobby, regardless of whether you live in a light-polluted city or in the dark countryside. Engaging even for beginners, astronomy can become an addictive pursuit—a scientific field where amateurs can still contribute and even make discoveries.

Of course, learning to navigate the heavens and tease out its hidden treasures requires time, patience, and, eventually, investing in equipment such as high-powered binoculars or a backyard telescope. Even with basic knowledge and equipment, sky-watchers can explore the wonders of the cosmos. But the sky really is the limit, with computer-tracking telescopes, photographic accessories—and perhaps even a backyard observatory. But that can wait. Let's start with the basics.

FURTHER

Backyard sky-watchers have found comets, asteroids, and exploding stars by taking advantage of modern technology. Exotic objects from distant black holes to exoplanets can be hunted down using digital imagers and robotic telescopes, and through online partnerships with professional scientists.

Exploring the Universe

Imagine gazing skyward on a clear night and seeing a flurry of shooting stars, or maybe peering through a telescope and spotting the moons of Jupiter or rings of Saturn. There are countless cosmic goodies ripe for the viewing if you know how, and where, to look for them. Whether you're watching the night sky from a downtown rooftop, a lounge chair in a suburban backyard, or a campsite far from city lights, this hobby never loses its luster.

Day or night, we can see through the thin atmospheric layer and spot many cosmic targets, some more obvious than others. During the day there's the sun, and maybe even the moon, moving slowly across the sky. When darkness falls, more distant stars begin to shine. While most of these stars are similar to our sun, they appear only as tiny points of light because of how far they are from us. A handful of them aren't stars at all but neighboring planets. The discerning sky-watcher can glimpse some of the nearest and largest—superhot Venus, desert-like Mars, monster-size Jupiter, and the ringed beauty Saturn—with nothing more than the unaided eye. You can even see ancient debris passing between these planets, like comets or meteors, which put on eye-catching, short-lived sky shows.

Look beyond our solar neighborhood and you will find a huge island of stars stretched out across the night sky: the Milky Way galaxy. This giant pinwheel-shaped island is home to approximately 250 billion stars, including our sun. In any direction, with the most powerful telescopes on Earth and in space, we see billions of other galaxies scattered across deep space in what we call the universe.

Exploring that universe requires nothing more than a walk outside and an open sky. The naked eye remains the most important piece of gear, and a focus on easily seen objects is the best way to build a foundation. Your first tasks as a sky-watcher may appear simple, but they will reveal much. Study the phases of the moon, track the motion of the stars and planets over weeks and months, and learn to recognize a handful of the brightest constellations visible in each season. Such observations will show you how the heavens work and allow you to build an appreciation for the fainter, more distant wonders you'll discover later.

THE SCIENCE OF Night Vision

Before sky-watching, 20th-century astronomer Clyde Tombaugh would prepare his eyes by sitting in the dark for an hour. When it's dark, the pupil will widen and open to let in more of the available light. Low- or no-light conditions allow the most light-sensitive cells in the peripheral part of your retina, called rods, to reactivate to their maximum level. If you don't have an hour like Tombaugh—who would later go on to discover Pluto—even 15 or 20 minutes in the dark will help. You can see details on even very faint celestial objects using a trick called averted vision. By looking out of the corner of your eye—instead of directly at an object—you will be able to discern much more detail, whether it's gaseous filaments in a nebula or the spiral arms of a galaxy.

SIZING UP THE GALAXY

THE MILKY WAY IS a spiral galaxy with gracefully curving arms and a bright, central bar of stars passing through its core, which holds a hidden black hole four million times the mass of the sun. To sustain life, planets embedded within the galaxy must avoid catastrophic threats such as close supernovae, gamma-ray bursts, and active black holes. They also can't be crowded in star clusters that would jostle them around too much. Luckily, Earth is in an ideal place for its inhabitants to thrive.

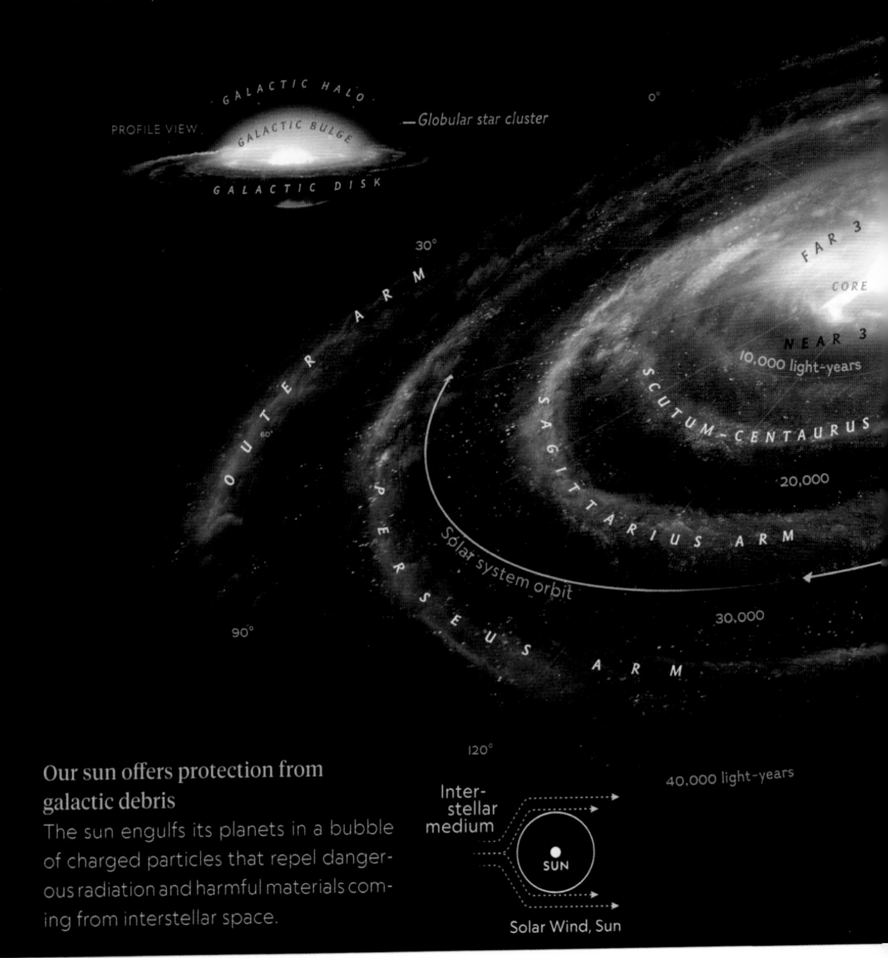

Our sun offers protection from galactic debris

The sun engulfs its planets in a bubble of charged particles that repel dangerous radiation and harmful materials coming from interstellar space.

Safe harbor

The Milky Way's arms are filled with hazards to habitability, including radioactive clouds, areas of active star formation, and sterilizing blasts from dying stars. Our solar system is nestled in a safe harbor between major spiral arms.

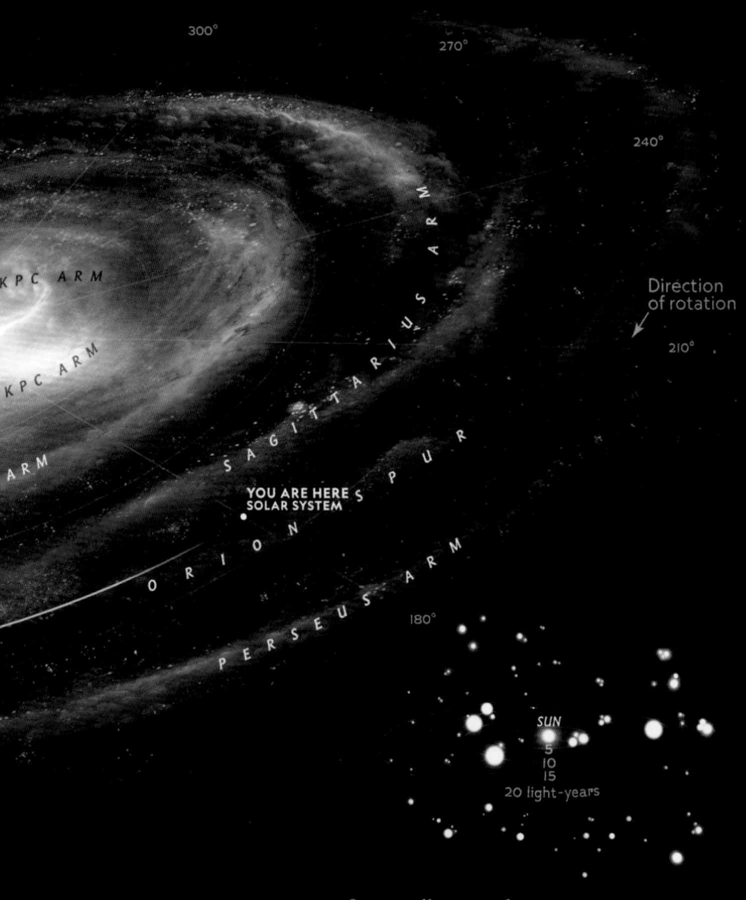

Far from stellar crowds

There are relatively few stars near the sun, reducing risks to Earth from gravitational tugs, gamma-ray bursts, or collapsing stars called supernovae.

GETTING ORIENTED

A time-lapse image of the sun, Venus, the moon, and Jupiter

REGARDLESS OF AN OBSERVER'S viewing location on Earth, all celestial objects appear to rise in the east, sweep across the sky, and set in the west in large, curving arcs. This daily motion is the result of our planet rotating around its axis once every 24 hours. Earth's continuous motion is what gives us night and day. And while daylight hides the stars from view, they continue to rotate around what are called celestial poles. Learning how to read and understand the celestial sphere is key to navigating the night sky.

FURTHER

Between the tug of the sun and the moon, Earth wobbles slightly as it rotates on its axis. This motion is called precession. The effect on celestial coordinates is so slight that backyard astronomers do not need to calculate for precession.

The Celestial Sphere

We know that Earth is not at the center of the cosmos, but star maps are still oriented that way. First, imagine Earth enveloped by a giant, distant globe. This globe is the celestial sphere. Its equator—the celestial equator—is parallel to Earth's, and lines extending out from Earth's North and South Poles pass through the north and south celestial poles. Imagine that all of the stars are attached to this sphere. Now set the celestial sphere spinning so stars rotate around the celestial poles. Earth actually revolves on its axis from west to east, but the effect is the same—stars on the celestial sphere pass overheard from east to west.

From the perspective of a viewer on Earth, half of the celestial sphere is always in view, arcing overhead like a hollow dome. If you were standing at the North Pole, you wouldn't be able to see the stars located south of the celestial equator. You would only be able to see so much of the sky before Earth got in the way, creating your horizon. Travel south, however, and more of the celestial sphere in the Southern Hemisphere will come into view. It is critical to know both your approximate latitude and the reference latitude of any star map you're using. Charts in this book are based on a view from latitudes close to 40 degrees north, the parallel that runs near New York City; Columbus, Ohio; Denver, Colorado; Salt Lake City, Utah; and northern California.

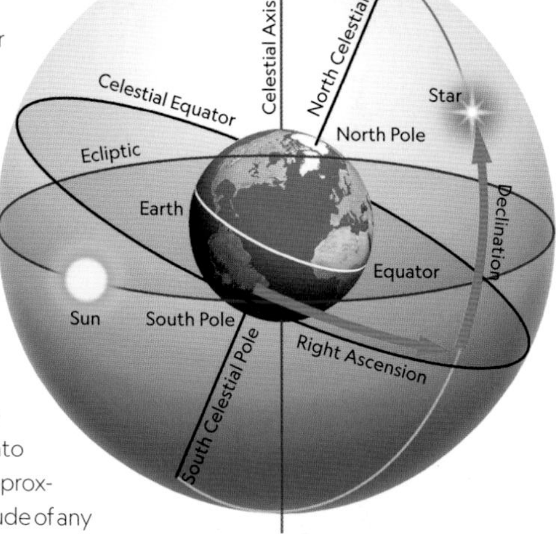

Coordinates of the Stars

Just like on terrestrial maps, where we use latitude and longitude, the sky has its own coordinate system projected onto the celestial sphere. This grid allows stargazers to identify any unique point in the sky. Declination is akin to latitude and marks how far the star is above or below the celestial equator. Just like on Earth it's measured in degrees, minutes, and seconds, with 0 degrees at the equator and 90 degrees at the poles. Positive declinations (+) are in the northern hemisphere and negative declinations (–) are in the southern hemisphere.

Right ascension, or RA, is akin to longitude and measured in hours, minutes, and seconds. It is divided, just like Earth's day, into 24 hours. The starting point, or "zero hour," is a north-south line that marks where the sun hits our Equator on the vernal equinox. Twenty-four hours is equivalent to 360 degrees, so each hour is 15 degrees of arc in the sky. For example, Polaris, or the North Star, has a declination of +89°, 15 minutes, and its RA is 2 hours, 31 minutes.

SKY-WATCHERS
Ptolemy

Like many early thinkers, second-century Greco-Roman astronomer Claudius Ptolemy was deceived by the apparent motion of the moon, sun, and stars into thinking that Earth was at the center of it all. He even constructed an Earth-centered model of the universe that accounted for the motion of the planets quite accurately. Each planet moved on an epicycle—a small orbital circle—that moved around a larger circle. He also had to assume that Earth was slightly tilted on its axis, something we know holds true even in our sun-centered reality.

THE ECLIPTIC

FURTHER

The apparent motion of the sun and stars provided one of the earliest ways to tell time, with the journey of the sun giving a sense of the day's progress and, at least in the Northern Hemisphere, the rotation of the Little Dipper around the North Star serving the same role at night.

ONE OF THE OBJECTS on the celestial sphere is the sun, which appears to follow its own course as that imaginary globe revolves around us. The apparent path of the sun is called the ecliptic—a line that reflects Earth's orbit around the closest star. The ecliptic determines much of what is visible to an earthbound observer. With a few exceptions at sunrise and sunset, the celestial objects overhead during daylight hours (and remember, the stars are always out there) are invisible to the unaided eye owing to the sun's brightness.

Celestial Band

The ecliptic is the major plane of the solar system, where all of the planets including Earth are embedded. A swirl of dust and gases formed around the nascent sun, all spinning in one direction in a vast flattened disk, and the material in the disk coalesced into larger bodies, which became the planets. From our vantage point, the ecliptic is like a planetary highway, a belt or band in the sky. Some planets fall a

Planets, a bright star, and the moon roughly align on the ecliptic.

On a "white night," the sun dips low but does not set.

little below or above the sun's precise ecliptic path, due to their own tilted orbits, but they are always nearby.

Days on Earth

Time itself is an astronomical construct. Technically, a "day" on Earth is the time it takes for the planet to turn once on its axis. That is measured in reference to the sun's passage from one meridian on one day, and its return to the same meridian on the next. Because the planet's orbital speed varies over the course of the year, the 24-hour clock is actually an average of these "solar days," and is referred to as mean solar time.

That differs from the time it takes for Earth to rotate once relative to the distant stars. A sidereal day, or star day, is about 23 hours and 56 minutes. Measured on a standard Earth wristwatch, in other words, a star that passes a given meridian at 9 p.m. one night will return there at about 8:56 p.m. on the next. That happens because our daily change in position relative to the Sun, though slight, is far greater than it is in reference to stars that are millions of light-years away. Because Earth is moving along its ecliptic path as it is turning, the planet must turn a bit farther than once around—about four minutes' worth—for the sun to pass the same reference point.

THE SCIENCE OF White Nights

St. Petersburg, Russia, may conjure dreary images of cold and dark, but from early June to early July the city celebrates the near-constant daylight—or "white nights"—that sets in around the summer solstice. Because of its location at nearly 60 degrees north latitude, St. Petersburg is pointed sharply toward the sun during the summer. The sun, in consequence, never drops more than a couple of degrees below the horizon, merely creating a sense of twilight until it rises the next day. At the North Pole the sun is in the sky from the time it rises on the first day of spring until it sets on the last day of summer.

CHANGING VIEWS

RECOGNIZING THE BRIGHTER STARS and constellations can give sky-watchers a clear sense of passing seasons, following along as their position changes in the sky. The other variable that factors into the visibility of particular stars on any given night is the position of the observer.

Following the Seasons

What we can see of the heavens changes with the time of night and the season. As Earth makes its 365-day orbit around the sun, the night side of our planet appears to face different constellations and each nightfall brings a slightly different patch of the "sphere." Stars invisible in June, when they are washed out by the noonday sun, will be riding in the night sky six months later when Earth has moved to the other side of its orbit. Star charts are typically monthly or seasonal to take this changing view into account and focus on the constellations best viewed at mid-evening.

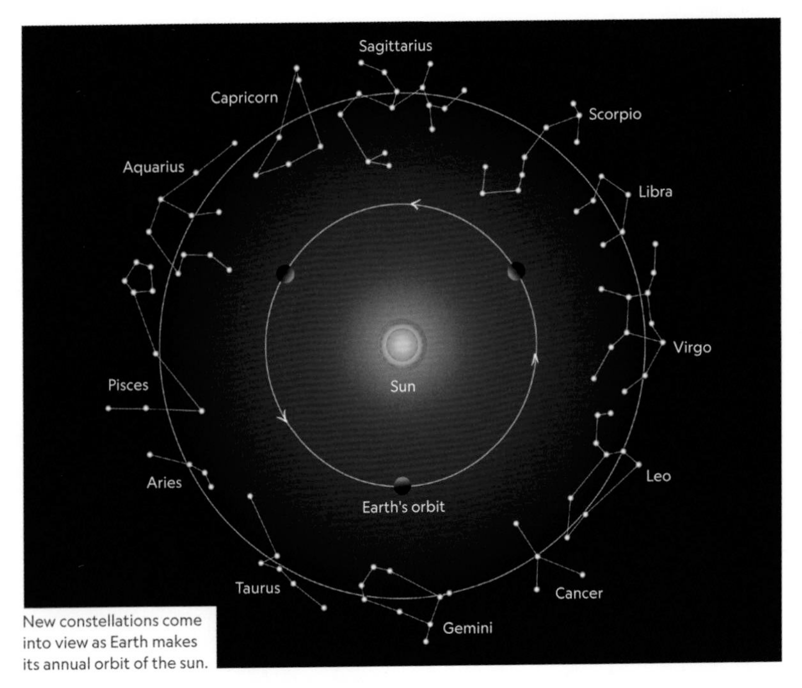

New constellations come into view as Earth makes its annual orbit of the sun.

A time-lapse photo shows how stars wheel around the north celestial pole.

View the sky at the same time each evening and new constellations will appear and disappear. For Northern Hemisphere observers, Orion rises in the east during November, rides high in the southern evening sky, heads westward in January, and by April is setting in the west at dusk. Meanwhile, eager stargazers can look at the predawn sky and get a sneak peek of the constellations that will soon be dominating the evening.

Changing Horizons

At different latitudes, the north celestial pole will lie at different altitudes above the horizon, and once you cross the equator, it will fall out of view entirely. At the North Pole, the north celestial pole will be straight overhead (90°) at what is called the zenith, with all other stars appearing to glide counterclockwise around it.

Farther south, at mid-northern latitudes like New York, Rome, and Tokyo, stars rise in the east at dusk and set in the west at dawn. Constellations like Ursa Minor and Cassiopeia still appear to never dip below the local horizon. Instead, these circumpolar constellations turn in circles every 24 hours without rising or setting. Stars near the south celestial pole remain invisible to observers in the Northern Hemisphere, since they never rise above the horizon. From the Equator, both celestial poles lie exactly on the north and south horizons (0°). The stars appear to rise in the east, move directly up the sky, and fall straight down in the west.

FURTHER

Stars that are connected by imaginary lines and form familiar stick-like figures have appeared the same across recorded human history. However, stars are not on a fixed plane, but rather lie at different distances from us and each other, so these patterns are actually fleeting on the longer cosmic timescales. For instance, in 100,000 years the bright stars that mark Leo and the Big Dipper will have moved sufficiently to make these patterns unrecognizable.

SKY MEASURES

FINDING YOUR WAY AROUND the night sky can seem confusing at first, a lot like learning how to read a map when visiting unfamiliar countries and cities.

Hand Over Fist

The apparent sizes of objects, and the distances between them, are measured in angles: degrees, minutes, and seconds. For example, the distance between the horizon to the point on the celestial sphere just above your head is 90 degrees. But it can be tricky to translate these measurements from handheld star charts to the sky above. An easy trick is to use your hands and fingers, held at arm's length, and the famous Big Dipper in the northern sky as a convenient angle-measurer.

Your outstretched hand is about 25 degrees wide from the tip of the thumb to tip of the little finger. This is roughly equal to the distance between the last star in the handle of the Big Dipper and the end stars in the bowl. Smaller distances can be measured with your fist, which is about 10 degrees across and is about equal to the width of the Dipper's bowl, while your three middle fingers measure about five degrees across, or about the same as the height of the Dipper's bowl. Your thumb is equal to two degrees and your index finger about a half degree, which would easily cover up either the sun or moon. We know that the sun's diameter is about 400 times larger, but the moon is much closer, which is why they have the same "apparent size" to an observer on Earth.

Finding Your Way

The constellations on the sky charts in Chapter 10 of this

FURTHER

Whether using a telescope or the naked eye, astronomers use bright signposts as a sort of road map that points the way to fainter objects, using their extended hands to approximate the angular measurement between the two objects. The Andromeda group of stars, for example, provide a stepping-stone to spotting the Andromeda galaxy, one of the closest galactic neighbors to our Milky Way galaxy and the deepest (farthest) object visible to the unaided eye.

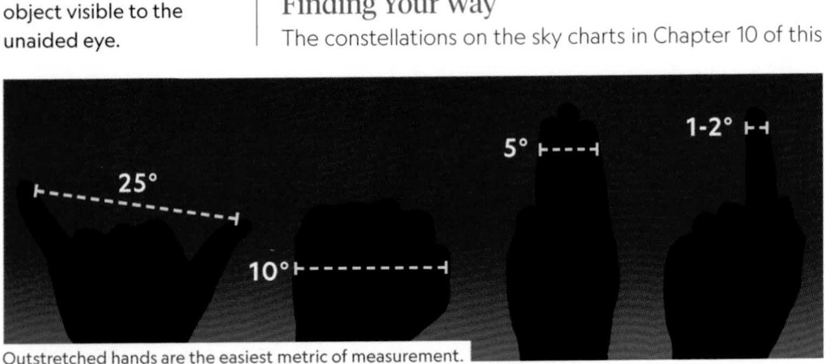

Outstretched hands are the easiest metric of measurement.

guidebook are all measured using this system of measurement. For example, Pegasus is about the size of two outspread hands, while the smaller Leo Minor is the width of a closed first. Practiced stargazers know that Orion's belt is about three degrees wide and the twin stars of Gemini (Pollux and Castor) are four degrees apart.

Angular measurement also helps stargazers determine their own latitude—a key factor in positioning your point of observation relative to the celestial sphere. Locate Polaris in the night sky and then use outstretched hands to measure the distance to the horizon. This angle is your latitude in degrees.

The Smallest Measurements

A single degree of distance can be sliced even smaller, into units called arc minutes or arc seconds. One degree is divided into 60 arc minutes, and each of those has 60 arc seconds. An arc second is about 1/1,800 the diameter of the moon as seen from Earth. Stars may appear to have a fixed position relative to each other in the sky, but they do travel in space. This movement is called proper motion and is also measured in arc seconds per year. For example, Barnard star's proper motion is 10 arc seconds per year—so it moves the width of the moon's disk in 180 years.

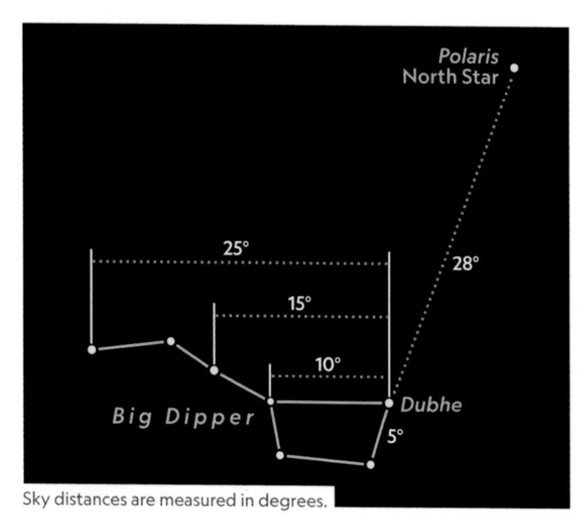

Sky distances are measured in degrees.

FURTHER

As Polynesian sailors settled across a massive geographic area beginning perhaps 3,000 years ago, they relied on an encyclopedic mental map of islands and the stars to guide them as well as cloud patterns and birds. The horizon was divided into sections based on the rising and setting of bright stars like Antares and Vega. Knowing that a certain island lay on a line between the rising point of one star and the setting point of another allowed for accurate navigation. These incredible voyagers likely traveled as far east as Chile.

STAR NAMES AND BRIGHTNESS

Northern constellations from a 1708 star atlas

FURTHER

Some companies have put the stars up for sale to whoever wants to name one. The IAU has disavowed such purchased star names, noting that "like true love and many other of the best things in human life, the beauty of the night sky is not for sale."

DIFFERENT CULTURES, different perspectives, different names: Polynesian seafarers created their own labels and legends for the stars, as did Native Americans and the Greek and Arab astronomers who gave us many of the names in common use today. What awaits any stargazing amateur is a hodgepodge of names—hundreds handed down from the ancients, thousands compiled in catalogs as telescopes grew more powerful, and millions assigned by scientific convention during the space age.

Naming Stars and Constellations

Though the International Astronomical Union (IAU) has established guidelines for several different catalogs of star names (which, for order's sake, have become strings of letters and numbers as opposed to elegant names like Vega), amateur observers need only be familiar with the common names and a few modern naming conventions.

Many civilizations have recognized the same star patterns and given them names and mythology. In the West, the 12 major constellations of the zodiac, and three dozen others, were handed down over the centuries, indexed most notably in Greco-Roman astronomer Ptolemy's *Almagest*.

Several hundred of the brighter stars also acquired proper names over the centuries. Arab astronomers provided much of that lexicon—from Acamar to Zuben Eschamali. Stars sometimes carry additional names that refer to their relative brightness in a constellation: Betelgeuse ("shoulder of the giant" in Arabic) is also Alpha Orionis (the brightest star in Orion).

Star Brightness

Gaze up at the night sky and you can see that not all stars shine the same. More than 2,000 years ago, Greek astronomer Hipparchus created the first stellar catalog that divided stars into different classes of brightness, or magnitude. The brighter a star appears (called its apparent magnitude), the lower its magnitude number. Ancient astronomers started with a magnitude 1 for the brightest stellar object and 2 for the next brightest, but this system did not account for even brighter stars like Sirius, planets like Venus and Jupiter, and the moon and sun, all of which receive negative numbers on the modern scale. The scale has even been extended to include faint objects only visible through telescopes. A star's magnitude is represented on star charts with a graded series of dots: The fainter the star, the tinier its "dot" is, with each dot size representing a whole order of magnitude in brightness.

Vision Limits

City sky-watchers can see stars down to around magnitude 4 on a clear, moonless night. From a dark location, far from city lights, the unaided eye can see stars down to magnitude 6. Binoculars will reveal stars as faint as magnitude 8 or 9, while small backyard telescopes can reach magnitude 12. A star's brightness depends on both its size and distance from us. Large stars inherently burn brighter than small ones, and stars located closer to Earth appear brighter.

FURTHER

The Hubble Space Telescope, in orbit above Earth, can record objects as faint as magnitude 30. That's about two billion times less bright than the faintest object visible to the naked eye.

SKY-WATCHING IN THE CITY

FURTHER

Visiting Dark Sky Parks offers the best possible views of the night sky and opens up countless deep-sky wonders. The International Dark-Sky Association (IDA) has a rapidly expanding list of such parks, including Grand Canyon National Park in Arizona, Cherry Springs State Park in Pennsylvania, Canyonlands National Park in Utah, Headlands in Michigan, and Big Cypress National Preserve in Florida. For a complete, updated list, visit the IDA website.

BETWEEN THE BUILDINGS BLOCKING the horizon and light pollution, urban sky-watchers may find it a bit more challenging. Deep space's fainter objects and low-horizon comets and stars might be out of reach, but there is still plenty to explore—particularly for the novice.

Urban Astronomy

The sun and moon, for example, are the two most obvious sky objects, and urban residents are at no special disadvantage when studying their movements. Observing these two bodies will teach you the basics of the solar system. The moon—big, close, and well-lit—will cut through urban glare and offer a rich study when it comes to its surface geography, phases, and eclipses. Beyond that, Venus reaches magnitudes as bright as −4.2, while Jupiter and Mars shine as brightly as −2.9 and −2.8, respectively, all well within the limits of city viewing. Mercury can be bright but its orbit close to the sun limits when it can be spotted. Saturn, with a magnitude of roughly 0.7, can also be seen with the unaided eye even in bright conditions.

Another 16 stars have a magnitude of 1 or less. They will stand out to city dwellers in their solitude, making

The same sky viewed from a city center, a suburb, a rural location, and a mountain observatory

them easy to use as star-hopping guides when you begin to acquire equipment for penetrating darker patches of the sky. Even in the city, binoculars and telescopes can bring thousands of stars into view—if you set yourself up correctly.

Light Pollution

From rural locations in North America, there may be as many 2,000 stars visible to the naked eye. However, anyone who has looked out the window of a plane at night has seen urban and suburban centers as a blaze of light, and, consequently, the best conditions for such sky-watching can be hard to find in those areas. Artificial light pollution comes from everywhere: lamps along highways and ever burning fixtures in parking lots and shopping malls, all of which reduce the number of visible stars. Light pollution can cut the viewing capacity of the human eye by a factor of around 40, from magnitude 6 down to 2.

There are ways to battle light pollution. First, deal with the light you can control. When observing from your own yard, turn off porch and interior lights, and ask the neighbors to do the same. If that fails, local parks may offer a refuge. Be strategic in where you point your gaze: Urban horizons tend to be bright and cluttered, while dark patches of sky high overhead make celestial objects much easier to see.

Dark Sky Preserves

Much of a city's light is wasted, cast up uselessly into the sky or spread far from the area it's supposed to illuminate. Shielded lights can serve the same purpose, and at a lower wattage, and several lobbying groups have argued in favor of more efficient lighting ordinances in order to preserve pristine night skies and help nocturnal migrating birds.

The main voice for light pollution abatement and protection is the IDA, which certifies public parks and privately owned spaces as Dark Sky Reserves and Dark Sky Parks. Dark Sky Reserves are usually near towns, but light pollution is being actively cut back. Dark Sky Parks have exceptionally starry skies and are protected specifically for the quality of their dark nights.

FURTHER

City-based stargazers can improve the view by turning off as many lights as possible. Avoid unshielded streetlights or security lamps. A black cloth held over your head can help block some of the invading light. Give your eyes time to adapt to the dark—a minimum of 15 to 20 minutes. Since you'll need light to read the star charts in this guidebook, tape a piece of red cellophane over your flashlight or buy a red-lensed light. Red light, especially when it's dim, hinders night vision much less than does white light.

SKY-WATCHING TOOLS

Binoculars with solar-observing filters

FAMILIARIZING YOURSELF with the major constellations using your unaided eyes will help you graduate more easily to magnified views through binoculars and telescopes. Naked-eye stargazing permits you to observe the entire sky, while most binoculars provide much narrower fields of view, equal to the width of 10 full moons. In comparison, telescopes give views about the size of only one full moon, making sky orientation more of a challenge.

Binoculars

Binoculars will nicely bridge the gap between the naked eye and the telescope and are truly an essential stargazing tool. Extremely portable, they provide a low-cost way to explore the starry skies. Finer models that produce pinpoint star images tend to range from $150 to $200. They can offer splendid views of dozens of craters on the moon, hundreds of star clusters, and even the four largest moons of Jupiter. The hazy band of the Milky Way comes alive with colors and structures that can be resolved into countless numbers of stars. Optical quality, magnification, and light-gathering power are paramount when choosing binoculars. For general viewing, a 7 × 50 configuration is a good place to start. The first number refers to magnifying power and the second refers to the aperture, or diameter, of the front lenses in millimeters. The higher the magnifying power, the greater magnification of space objects, though it also magnifies the effects of shaky hands.

FURTHER

Though hardly exotic, binoculars bring a wealth of objects into view:

- Comets
- Large craters on the moon
- Jupiter's four biggest moons
- Mercury, Uranus, and Neptune
- Star fields along the Milky Way
- Andromeda and the Pinwheel galaxy

Telescopes

With a good working knowledge of the sky and the right telescope, the beauty of the sky will unfold before you. Well-crafted amateur telescopes range in price from $200 to $5,000 (or more) and fall into two main types: lens-based refractors and mirror-based reflectors. Avoid telescopes advertised by their maximum power and commonly found at department stores. The main features to look for are quality optics and a rock-solid mount. Look for one with a structure made up of metal and wood, not plastic, and avoid models where the image wobbles every time you touch the scope. Keep in mind that the larger the primary mirror or objective lens is, the brighter and sharper the images will be.

Software

Backyard stargazing has undergone a revolution in less than two decades thanks to the rapid pace of technological advances. Planetarium programs and robotic motor drives available on many backyard telescopes come loaded with thousands, or even millions, of objects in their databases. Select your object from the list and the scope will find it.

Planetarium apps such as Sky Safari and Stellarium available for download on computers, handheld tablets, and smartphones show millions of stars and realistic photographs of objects and can even guide naked eye observers to specific sky targets. The Sky Chart app will work on a handheld device.

FURTHER

Keep in mind that larger binoculars become heavier and harder to hold as you use them. Indeed, the biggest problem is holding them steady, so try to rest your arms on something so the stars don't dance like fireflies! Even better, mount the binoculars onto a camera tripod with special mounting brackets, available at most outdoor stores. You will notice a huge difference in image clarity, making your search for star clusters and double stars a breeze.

Tablet and smartphone apps guide observation.

STARGAZING AS A FAMILY

THE SKY IS ALIVE with myths and monsters and gods, whether it's the legends behind the constellations, the names of the planets, or the wide variety of cultural tales associated with phenomena like the aurora borealis. Science fiction, meanwhile, brings us the promise of a visit by a friendly extraterrestrial or a menacing Martian attack launched à la H. G. Wells.

For kids, at least at first, stargazing may be less about the science and more about the enduring mystery of our place in the cosmos, and the ancient stories. To a young mind, the broad brush of a dark sky may be more intuitively satisfying than the hunt for a planet or distant galaxy.

That sense of imagination can make for some wonderful family outings, but it may involve making choices regarding the type of equipment being used and the amount of time spent in the field. While most children may not go on to be astronomers, some will find this hobby inspires them to follow a career path in science, technology, engineering, or math. But even if their interest is fleeting, an appreciation of nature and the wonder of a starry sky will last a lifetime. For that, no fancy equipment is needed.

A Kid's-Eye View

Binoculars are a particularly accommodating way to look at the sky with children.

Telescopes bring deep-sky treasures into view.

They require no setup and, being less expensive than a telescope, will cause fewer worries about accidents. If you have a choice, opt for a smaller, lighter size to better suit kids' hands. With a wider field of view, binoculars bring in more sky and may be best suited to the things a child wants to see and the way he or she wants to look at them. Have at least one pair (or more) on hand to share.

When it comes to viewing through telescopes, remember to keep young children's expectations realistic about what they might see. Outside of the moon, perhaps Saturn and Jupiter, and a few star clusters, most celestial objects won't look like the photos in books and on the Internet. The human eye is not sensitive enough to pick up the colors in a nebula, but we can glimpse the delicate wisps and tendrils in their cloud structures. With older kids it's a great idea to show pictures of selected objects, explain what they are and their distance in space and time, before observing them. Doing so will give them a deeper appreciation of the wonder of being able to glimpse distant cosmic objects with their own eyes.

Picking the Right Time

Timing a skygazing trip with kids can be tricky. The late-setting sun in summer may leave you with little time before they begin to get tired. Go before sunset and plan the outing around appropriate objects—a first quarter moon, perhaps, or a planet when it is near opposition. When the timing's right, see who can spot Venus first. Winter offers an earlier night and new star patterns, but kids are more susceptible to cold than adults. Make sure to bring an extra layer or two of clothing (and a thermos of hot chocolate) to make sure everyone is comfortable.

For a child who is serious about the hobby and ready for a telescope, consider some of the easier-to-use models, particularly those with Dobsonian or computer-driven mounts (described on page 136). Ideally, you'll want to invest in something the child can independently set up and use.

Last—and most important for any viewing when the sun is still in the sky—impress on children that the sun can seriously damage their eyesight, and be on guard with your equipment to be sure no one hazards a look through the unfiltered lens.

SKY-WATCHING PROJECTS FOR THE FAMILY

SATELLITES

TIMING: Year-round, just after sunset or just before sunrise

WHAT TO LOOK FOR: Starlike lights that quickly and steadily move across the sky (see page 80)

VENUS

TIMING: Check almanacs and guides to see if Venus is best seen in the morning or evening.

WHAT TO LOOK FOR: A steady, bright orb that does not twinkle like a star (see page 94)

BIG DIPPER

TIMING: Year-round

WHAT TO LOOK FOR: The most easily recognizable star pattern (see page 196)

SIRIUS

TIMING: January and February

WHAT TO LOOK FOR: The brightest star in the sky, which belongs to the constellation Canis Major (see page 269)

PERSEIDS

TIMING: July and August

WHAT TO LOOK FOR: Bright "shooting stars" that originate near the constellation Perseus (see page 133)

Clouds and the thin blue line of Earth's atmosphere viewed from space

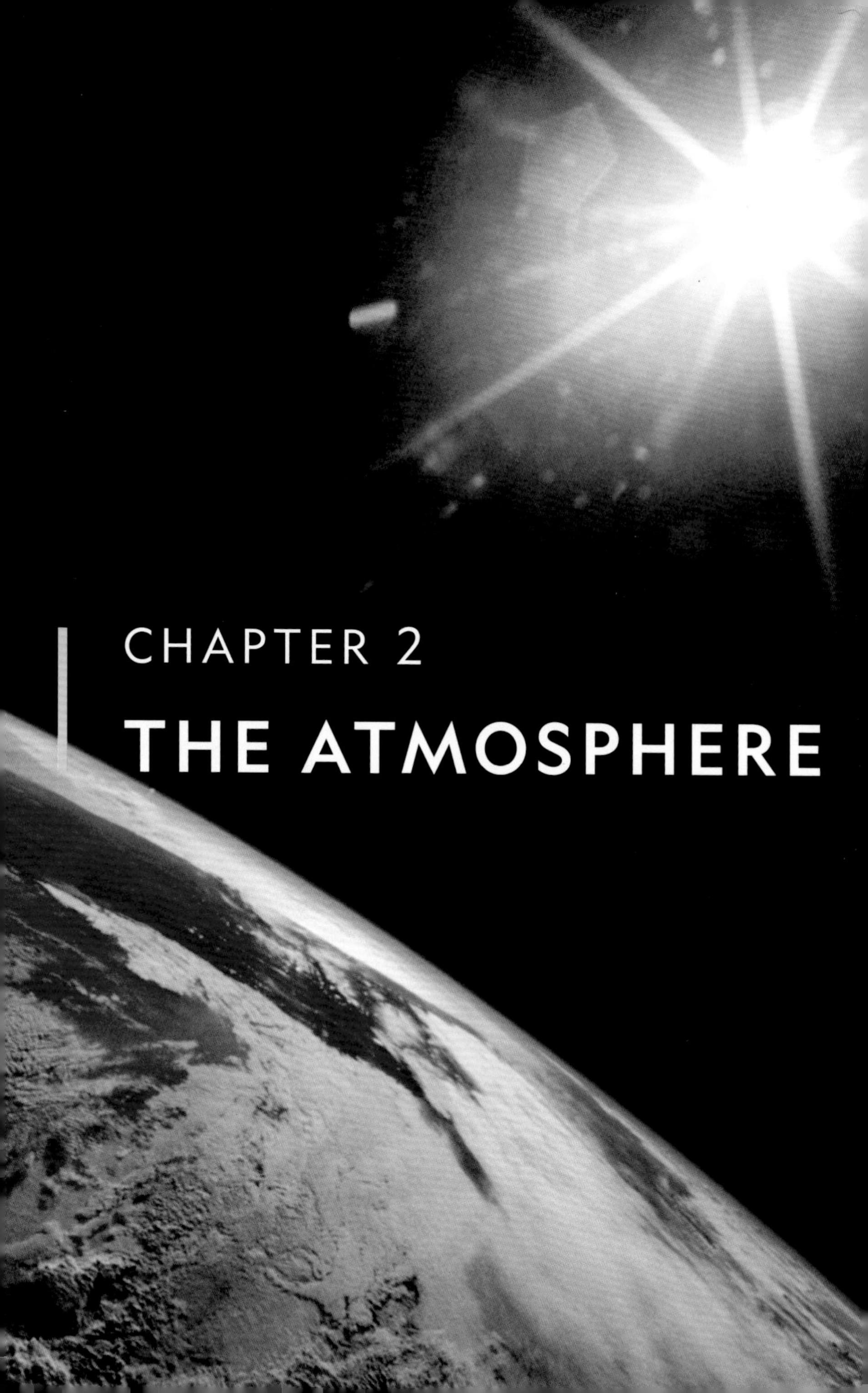

CHAPTER 2

THE ATMOSPHERE

EARTH'S THIN SKIN

LIGHT FROM THE SUN and stars travels immense distances to reach Earth, yet the last few dozen miles of the journey are what have the greatest effect on what you'll see. After crossing the comparatively empty reaches of interstellar space, the small portion of a star's light that does reach Earth must still navigate the soup of gases, liquids, and solids that make up the planet's atmosphere, which has played tricks on sky-watchers since the very beginning of recorded time. As a practical matter, the atmosphere is only a few miles deep, with about 98 percent of atmospheric mass contained within some 20 miles (32 km) of the planet's surface.

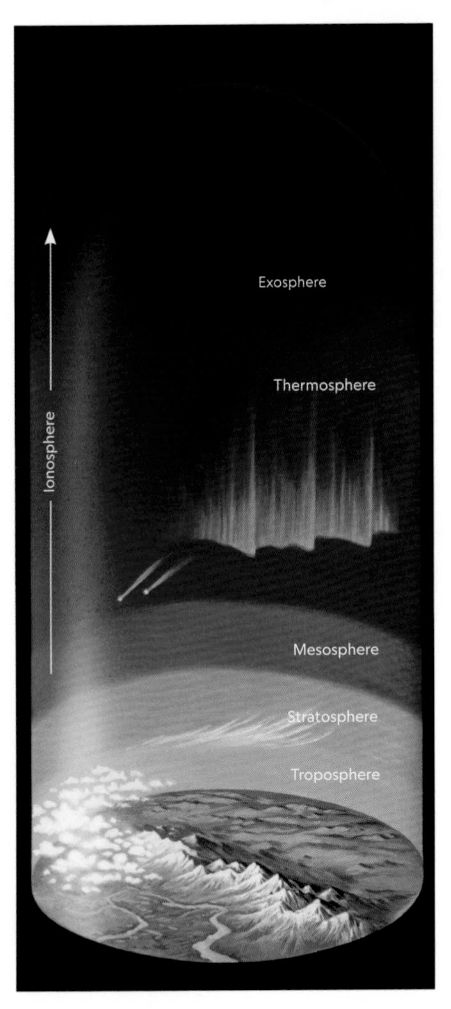

Inner Layers

At the lowest level, extending roughly 5 to 9 miles (8 to 14.5 km) out from Earth's surface, the troposphere has the biggest impact on visibility because weather "happens" here. Enveloping cloud systems in the troposphere can block the sky for days. Dust and pollution, particularly near cities, can overwhelm the light of all but the brightest objects.

Starting at the edge of the troposphere, the stratosphere is where upper-level ozone concentrates and blocks some of the more intense and penetrating forms of ultraviolet solar radiation. Clouds rarely appear in this layer; the few you'll see are thin wisps called nacreous or mother-of-pearl clouds.

Outer Layers

In the mesosphere, a band that runs between 31 and 53 miles (50 and 85 km) above Earth, temperatures drop as low as minus 120°F (−84°C). Cloud formation all but stops. Any clouds that do form—clusters of visible ice crystals—are rarely seen

Weather occurs in the troposphere.

and mostly at twilight in northern latitudes. Meteors burn up in this layer, appearing as shooting stars to lucky observers on Earth.

The thermosphere is the largest of Earth's atmospheric zones as well as the hottest. Though thinly distributed, the gases in this thick layer absorb so much arriving radiation that temperatures can exceed 3632°F (2000°C). The collision of solar rays and gases also produces electrically charged regions—contained in the ionosphere—that reflect radio waves back to Earth: a boon to modern telecommunications. Low-Earth-orbiting satellites including the International Space Station orbit in this layer and aurora borealis occur here. The thermosphere is where Earth's atmosphere ends and space truly begins.

The uppermost exosphere officially extends from 372 to 6,200 miles (600 to 10,000 km) with near vacuum conditions. However, evidence suggests it may reach out to 62,137 miles (100,000 km): that's a quarter of the way to the moon.

THE SCIENCE OF Observatories

More than a thousand years ago, Maya scientists used the raised platform of El Caracol in modern-day Belize to watch the skies. Centuries later, astronomers are still seeking the best views. The W. M. Keck Observatory, on the peak of Mauna Kea in Hawaii, is one of the world's highest observatories, peering from a point that is above about 40 percent of Earth's atmosphere and 97 percent of its water vapor.

LAYERS OF THE ATMOSPHERE

TROPOSPHERE
ALTITUDE: 5-9 mi (8-14.5 km)
ACTIVITY: Lowest level where clouds form and weather occurs

STRATOSPHERE
ALTITUDE: 9-31 mi (14.5-50 km)
ACTIVITY: Contains ozone layer that absorbs sun's ultraviolet radiation

MESOSPHERE
ALTITUDE: 31-53 mi (50-85 km)
ACTIVITY: Temperatures drop sharply to roughly −135°F (−93°C)

THERMOSPHERE
ALTITUDE: 53-372 mi (85-600 km)
ACTIVITY: Sun's radiation causes sharp increase in temperature, up to 3632°F (2000°C)

EXOSPHERE
ALTITUDE: 372-6,200 mi (600-10,000 km)
ACTIVITY: Outermost layer of atmosphere, mostly hydrogen and helium gas

NIGHTFALL

"Earthshine" illuminates the full orb during a crescent moon.

FURTHER

Sunlight contains every visible color; however, the sky appears blue because of Rayleigh scattering, or the angle at which sunlight is scattered off molecules in the atmosphere. Light from the blue part of the spectrum scatters more easily and dominates what we see during the day when sunlight's trip down to Earth is shortest. At sunset, the molecules in the air bend the sunlight toward the red part of the spectrum, which is what causes bright red sunsets.

THE SUN'S APPARENT MOTION across the sky fascinated and confused the world's finest thinkers for thousands of years. Twilight is a great time to pay attention to how our closest star, our planet, its moon and atmosphere all interact.

Twilight Phenomena

As the sun sets, a glance to the east can reveal a deep blue band across the horizon—the shadow of Earth cast onto the sky. In areas with high mountain peaks, a "summit shadow" may also appear. The band rises as the planet rotates, until night falls and the shadow becomes invisible. Under clear skies and with a broad horizon just before sunrise or after sunset, the pink Belt of Venus, caused by a sliver of sunlight reflecting off the atmosphere, can be seen atop Earth's shadow. Similarly, earthshine can make a crescent moon seem almost full. The moon shines because it reflects light from the sun, but also because Earth illuminates it. Our planet is covered with highly reflective water and white clouds, so we act as a gigantic reflector, bouncing sunlight onto the crescent moon and lighting up the portion of its disk that would otherwise be dark.

Nightfall can also reveal a subtle, pyramidal glow in the western sky, most visible at latitudes close to the Equator—and then only on dark, moonless nights. The pyramid of light near where the sun sets is zodiacal light, so called because it occurs along the ecliptic—the path that the sun and moon follow, and the approximate course of the constellations of the zodiac. Zodiacal light is created when sunlight reflects off dust in the plane of the planets' orbits.

Clarity and Transparency

Clear night skies may look inviting, but moving, turbulent air in the atmosphere can impact the view of distant stars, making them appear to shake and ripple as if they're being looked at through water. Starlight traveling through cold and hot layers of air can bend in different directions, resulting in blurred views called poor "astronomical seeing." This effect is only magnified by binoculars and telescopes. The light-bending power of air is diminished in flat geographical regions, where air can move smoothly across the land. Turbulent air predominates in the days after a weather front moves through.

Meanwhile, details of planets and distant galaxies can be washed out by smoke, pollution, and moisture and humidity, which all lower the transparency, or clarity, of the atmosphere for stargazers. Transparency tends to improve at higher altitudes and after rainstorms sweep away and clear the surrounding air mass of particulates.

THE SCIENCE OF Green Flashes

Most often seen during coastal sunsets, green flashes are rare but memorable. The atmosphere acts as a prism that bends sunlight into separate colors, each on its own wavelength, then disperses it. At sunset, the longer wavelengths of red, orange, and yellow disappear first, but the atmosphere disperses the other colors more slowly, so they disappear shortly after. On a very clear day with optimal conditions, the green light will still be evident for a few seconds just as the sun sinks below the horizon. The top of the sun will have a sliver of green that may stretch out in a brief green flash or glow.

Cerro Paranal Observatory in Chile's barren Atacama Desert offers pristine views.

TRICKS OF THE LIGHT

LIGHT FROM THE SUN is called white light. White light is composed of all different colors that the human eye can see: red, orange, yellow, green, blue, indigo, and violet. Hang a glass prism in a window with direct sunlight and the prism will refract a beam of rainbow colors. When you see a rainbow in the sky, the sunlight has been refracted through water droplets.

The play of sunlight during the day can produce some beautifully unique effects. Rainbows are fairly common, but lesser known and rarer perhaps are rainbows seen at night. Lacking the brilliant colors of their daytime cousins, lunar rainbows, or "moonbows," form when sunlight reflects off the moon's bright surface and then refracts off water droplets in Earth's upper atmosphere. Double moonbows are possible when there are rainstorms in the area, but the moon needs to be near full phase to provide enough light to form moonbows.

Light Pillars

Generally seen in winter and in high-latitude regions during the coldest months, vertical columns of light can

Reflective clouds create a moon halo.

Light pillars appear in Russia's icy atmosphere.

form above or below a light source when snow dust blows near ground level. Known as light pillars, these optical phenomena form around the sun, moon, and sometimes even artificial lights like streetlights, in which case the pillar takes on the color of the light source and can range from red, orange, yellow, green, and yellow.

Halos and Sun Dogs

Hexagonal ice crystals in high-flying cirrus clouds can also bend or refract light into stunning halos around the sun, or even the moon. These thinly veiled, frosty clouds form above 20,000 feet (6,000 m) and may be invisible to observers on the ground. While moonlight is too weak for much color to form, solar halos often appear red along the sharp inner edge of the ring and blue along the outer, more diffuse rim.

Sun dogs, also known as mock suns or parhelions, are glowing orbs on either side of the sun that, like halos, are created by sunlight refracting off ice crystals in cirrus clouds. Sun dogs, however, are very common and can be seen anywhere in the world where there are ice crystals in the upper atmosphere. These ghostly apparitions tend to be most visible 22 degrees away from the sun as it sets.

In rare cases, sun dogs can stretch across large portions of the sky, which has caused them to be mistaken for auroras. When it comes to the most common sun pillars, they are usually seen when the sun is near the horizon.

THE STORY OF Sun Dogs in History

The concentrated patches of light that appear on either side of the sun have been recorded by sky-watchers for thousands of years. The phenomenon's scientific name is parhelion, a Greek word meaning "beside the sun." The first recorded description was made by the ancient Greek philosopher Aristotle, who in the fourth century B.C. noted, "Two mock suns rose with the sun and followed it all through the day until sunset." In medieval times, sun dogs were considered omens. In one instance made famous by Shakespeare, sun dogs hovering above a battlefield in 1461 spurred on the troops loyal to the future king Edward IV, fighting to victory during England's Wars of the Roses.

EARTH'S MAGNETIC FIELD

THE SUN IS EARTH'S PRIMARY SOURCE of energy, but solar radiation arrives at the surface only after a long journey and much filtration—for which we can be thankful. The planet's atmosphere blocks many potentially damaging ultraviolet and other rays, while allowing just enough energy through to keep the planet's surface temperate. But long before that a bubble-like magnetic shield, sustained far beyond Earth's atmosphere by the planet's inherent magnetism, deflects an otherwise intense atomic bombardment.

SKY-WATCHERS
James Van Allen

James Van Allen's long-standing interest in cosmic rays led to a profound discovery. When the first U.S. satellite, Explorer 1, went into space in January 1958, it carried Geiger counters that Van Allen developed for the trip. A steady flow of evidence indicated that Earth was encircled by two doughnut-shaped belts where the planet's magnetic field was trapping solar and cosmic radiation. Considered the first major discovery of the space age, the belts were named for the scientist who discovered them.

The Magnetosphere

With a swirling interior of molten metal, Earth's core generates a magnetic field that resembles a conventional magnet but on a much larger scale. Current flows outward from the two poles and loops back toward the center. That fact has been of practical use since the invention of the compass. But the breadth and impact of Earth's magnetic field have only become fully clear in the satellite age, when scientists discovered the size and protective nature of a "magnetosphere" extending tens of thousands of miles into space. The magnetosphere deflects the steady stream of atomic particles known as solar wind, a bit of "space weather" that can still knock out communications and electronic equipment during intense geomagnetic storms.

Buffeted by this solar wind, the magnetosphere changes size and shape throughout the day. On the side of Earth facing the sun, the solar wind compresses the magnetosphere down to perhaps 40,000 miles (64,000 km) from the surface. On the planet's opposite side, the solar wind may "blow" the magnetosphere into a long tail of 50 to 100 times that length, thus extending far past the orbit of the moon.

James Van Allen

Van Allen Belts

Although the sheer strength of the magnetosphere repels some solar wind, atomic particles from the sun do push their way through, as does inter-

Earth's protective magnetosphere under the influence of solar wind

galactic radiation from outside the solar system. In the late 1950s, physicists found out where some of that atomic material goes.

Girding Earth are two concentric zones of intense radioactivity, dubbed the Van Allen belts, named after the physicist who discovered them in 1958. Thickest at the Equator and weakest at the poles, the inner belt begins approximately 600 miles (1,000 km) above the surface and extends outward about 3,000 miles (4,800 km). The outer belt runs from roughly 10,000 to 25,000 miles (16,000 to 40,000 km).

The Van Allen belts are composed of atomic particles—protons and electrons from outside the solar system and helium ions from the sun chief among them—that have pierced the outer rim of the magnetosphere, mingled with the atmosphere and, in effect, been caught in a magnetic spiral, bouncing from pole to pole. This accumulation of radioactive particles is so thick with energy that spacecraft navigating through them need protective shielding for their equipment.

THE SCIENCE OF Magnetic Flips

Earth's magnetic field has reversed its polarity many times over the past three billion years. Unlike what doomsday theorists may tell you, these flips are common and happen over many centuries—if not millennia. Geological records show that throughout each of these reversals, the positions of both poles continually shift and slide across the globe. Since the early 19th century the magnetic north pole has migrated northward by more than 600 miles (1,000 km) and continues to slide about 40 miles (64 km) per year.

SOLAR WIND AND AURORAS

SOLAR WIND INVOLVES a fairly constant outpouring from our sun—roughly a million tons of matter per second, streaming away at speeds as fast as 2 million miles per hour (3 million km/h). But that pales compared with the amount of material spilled out during events like coronal mass ejections and solar flares, phenomena that can dump billions of tons of charged plasma and particles into space.

FURTHER

Solar storms in 1989, at the peak of one 11-year cycle, caused auroras that were spotted as far south as the Caribbean.

Polar Light Show

From Earth, violent solar wind events are responsible for one of the most colorful and anticipated atmospheric phenomena—the dancing auroras that light up the north-

ern and southern sky. The heavier-than-usual flow of charged particles breaches the magnetosphere and reaches into Earth's atmosphere, coming perhaps as close to the surface as 50 miles (80 km). There, this solar material energizes molecules of oxygen and nitrogen in the atmosphere and causes them to glow like neon signs.

For viewers on Earth, the resulting curtains of green, red, and pink light are a dazzling spectacle, intensifying in relation to the solar storms that cause them. Data from satellites launched to study auroras and other electromagnetic activity explained why the lights seem to shimmer so vibrantly: As solar wind blows, it actually stretches part of Earth's magnetic field, which eventually snaps back into place. Energy released during that magnetic reconnection causes the aurora to brighten and undulate.

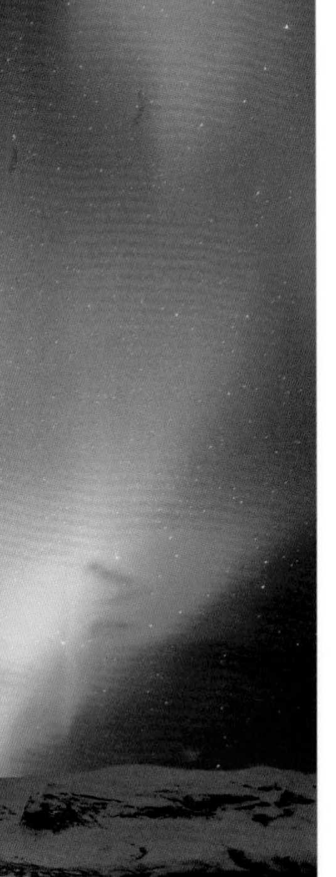
Majestic aurora above Iceland

North and South

For the most part auroras are polar phenomena, hence their formal names: aurora borealis, or the northern lights, and aurora australis, or the southern lights. They are often seen in the high latitudes of North America—Alaska, and certain parts of Canada—and in Scandinavia, where the inhabitants crafted many legends around the ghostly lights. In the south they shine mostly over remote areas of New Zealand and Tasmania or over oceans and sparsely populated Antarctica. However, when solar storms are particularly intense, they can be seen into the middle latitudes. Earth's magnetic field lets more solar particles interact with the atmosphere near the equinoxes, so auroras are particularly active in March and September.

THE SCIENCE OF STEVE

While eerie, dancing auroras have mesmerized skywatchers since prehistoric times, a new type of purplish sky glow has been discovered, fondly called STEVE (Strong Thermal Emission Velocity Enhancement). The phenomenon was first detected by amateur aurora-chasers who discovered evidence of a rare purple aurora that arcs across the spring and fall night skies for up to an hour. Scientists flew a satellite through STEVE and recorded temperatures of 5400°F (3000°C) and high speeds of movement. These physical properties match up with an atmospheric event called a subauroral ion drift: a rapid flow of charged particles interacting with Earth's magnetic field.

NOCTILUCENT CLOUDS

EVERY YEAR, as early summertime skies turn to deep twilight, reports ramp up in both hemispheres about eerie electric-blue clouds shimmering in nighttime skies in high-latitude regions known as noctilucent, or night-shining, clouds.

Sparkling Dust

These glowing, silver-blue wisps form around Earth's polar regions in the mesosphere, the highest level of Earth's atmosphere, far above where most clouds occur. Near the edge of space, around 50 miles (80 km) up, temperatures are a bone-chilling minus 148°F (−100°C), the air a million times drier than any desert. Under these extreme conditions, water vapor freezes onto any floating dust particles, seeding the ice crystals that form the tendrils and filaments of noctilucent clouds. Around dusk and dawn, the sun brings these clouds to life, making them glow against the dark backdrop of the twilight skies. When the sun is out of view at roughly 10 degrees below the horizon, light cannot reach low tropospheric clouds, but it does illuminate the area of the sky at high altitudes where noctilucent clouds form.

Noctilucent clouds were first recorded in 1885 after a volcanic eruption on the Indonesian island of Krakatoa, which sent a massive ash cloud into the upper atmosphere that circled Earth for months. Spectacular red sunsets and distinctive glowing clouds persisted for much longer. While such large volcanic eruptions are not all that frequent, nearly 100 tons (90.7 t) of meteoritic dust falls on Earth

Sunlight below the horizon illuminates high-hanging noctilucent clouds.

This image shows noctilucent clouds formed as streaky bands and whirls and as rippling waves.

every day, and this meteor smoke forms the foundation of noctilucent clouds.

How to See Them

Noctilucent clouds come in four different forms: veil, similar to bright fog; bands, parallel streaks; waves, the characteristic ripples; and whirls, large looped or twisting structures. They appear from May through August in the Northern Hemisphere, and from November through February in the Southern Hemisphere, when temperatures in the mesosphere are at their coldest. To catch sight of this beautiful seasonal phenomenon, look toward the northwest when the sun is below your horizon about an hour after sunset. You can also look for it in the mornings in the northeast about an hour before sunrise.

High latitude areas between 50° and 70° north or south of the Equator are best positioned to see the clouds—roughly north of Boston or Seattle. However, over the last century these unusual formations have been spotted more frequently and much farther south in places such as Utah, Kansas, and Colorado. While it's a mystery why they appear to be spreading, some scientists have suggested a link to climate change.

The clouds are also visible from space—astronauts on board the International Space Station have reported seeing them, even capturing the occasional image. Similar clouds were even spotted on Mars back in 2006, when the Mars Express orbiter saw them floating some 60 miles (100 km) up from the planet's surface. These Mars clouds are likely formed by frozen carbon dioxide.

SPRITES AND ROCKET TRAILS

FOR MORE THAN TWO decades, keen-eyed sky-watchers in the right place at the right time have been reporting and capturing mysterious reddish orange flashes of light high in the night skies, dubbed "sprites." Considered a myth until an airline pilot flying over the United States filmed their flickering lights above a thunderstorm in 1989, these momentary bursts of electricity can literally reach the edge of space, about 50 miles (80 km) above the planet's surface.

Rocket trail from a SpaceX launch

They have remained elusive because they last for only a few milliseconds and are normally hidden by the clouds. Sprites are related to lightning; however, this rare, bizarre phenomenon is produced when discharged electricity shoots out from the top of a cloud to the edge of space instead of heading to the ground.

Chasing Sprites

You have the best chances of seeing sprites throughout the Midwest United States from Colorado to Minnesota, and as far south as Texas. Around the world, sprites have been seen in storms above South America, Africa, and Australia. To see them with the naked eye during a storm, find a sheltered location far away from the blinding lights of the city. Haze and air pollution can also block sprites from view. Gaze well above the top of a thundercloud while blocking out all the lightning action below with a piece of cardboard. Expect them to occur every 10 minutes or so on average at the height of the storm. Expert sprite chasers suggest, as a first step, to check the Internet weather services for strong thunderstorms within 500 miles (800 km) of a location using regional radar maps. Although astronauts have the perfect vantage point looking down from space, ground-based viewers can also get a height advantage by placing themselves on mountainsides overlooking storms moving across the flat plains below.

Rocket Trails

There is nothing quite like watching a rocket launch from Earth and into orbit. Only a relatively lucky few get a chance to watch up close that moment when it leaves the launchpad. Millions, however, can witness its steam exhaust form hypnotic, narrow clouds that twist and drift as it reaches many miles into the heavens.

These exhaust trails form as rocket fuel mixes and burns in the combustion chamber during a rocket's booster stage. The resulting pressure from the explosion exits the engine nozzles, pushing the rocket skyward. Hot water vapor exits the engines at high speeds and, much like with high-flying jet planes, condenses quickly into what is called a condensation trail, or contrail for short, in the much colder air at high altitudes. The resulting, spectacular contrails are visible for hundreds of miles.

As they reach the highest, thinnest atmosphere layers, they not only condense but freeze and expand, and winds can cause them to spread into strange patterns. These twisting shapes are accentuated when launch occurs near sunset, when contrails become particularly reflective. Within an hour after sunset or an hour before sunrise, the sunlight is below the horizon but it hits the contrails at steep angles, so they shine against the dark sky.

With the increase of commercial ventures into space, we can expect many more opportunities to view contrails in the years to come.

A red sprite is visible above an active lightning storm and below the rising moon.

Life on Earth depends on the energy, light, and warmth of the sun.

CHAPTER 3

THE SUN

OUR STAR

SYMBOL: ☉

AVERAGE SURFACE TEMPERATURE: 10,000°F (5500°C)

AVERAGE CORE TEMPERATURE: 27,000,000°F (15,000,000°C)

ROTATION: 25.4 days at equator

DIAMETER: 924,000 mi (1,500,000 km) at equator

MASS: 333,000 times that of Earth

GRAVITY: 28 times that of Earth

ROUGHLY 4.6 BILLION YEARS AGO, in the Orion arm of the Milky Way, a spinning cloud of hydrogen and other interstellar matter succumbed to the effects of gravity and began collapsing on itself. As the cloud of gas condensed, pressure and temperature at the center increased so dramatically that the hydrogen atoms began fusing into helium, releasing immense amounts of energy in the process. Radiating outward, the energy helped counteract gravity and stop the cloud's contraction. Equilibrium set in and a star was born—our star, the sun.

Situated about 93 million miles (150 million km) from Earth at the center of what is, since Pluto's demotion, an eight-planet solar system, the sun is a daily reminder of our own tenuous place in the universe. It is one of hundreds of billions of stars in the Milky Way, which is itself one of per-

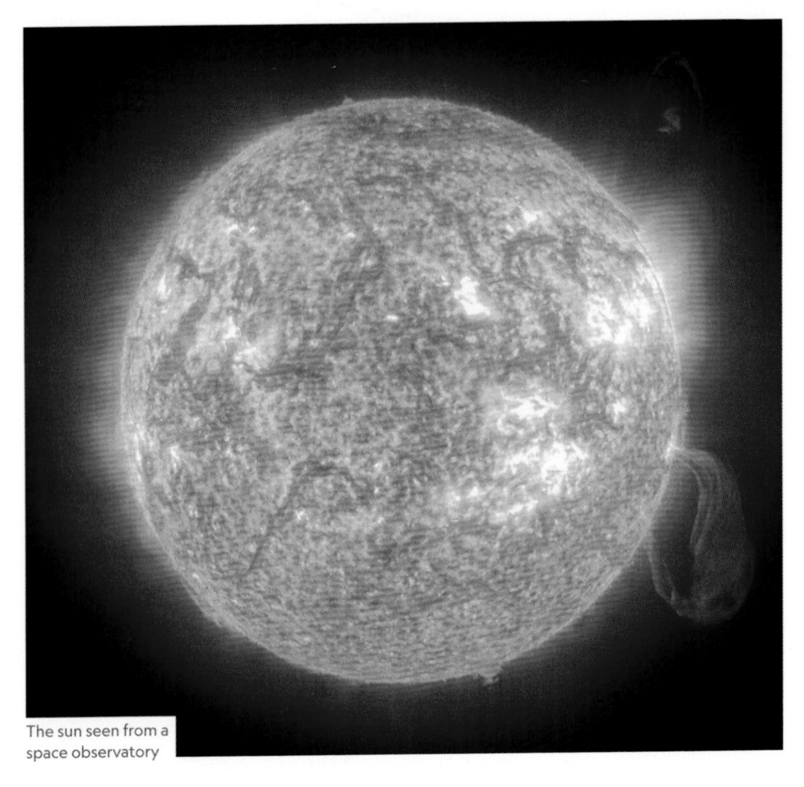

The sun seen from a space observatory

haps hundreds of billions of galaxies in the universe, and its dimensions, dynamics, and distance from Earth provide enough energy and heat to support life without overwhelming the planet with violent radiation. Had the original gas cloud been larger or smaller, or the influence of other forces different, Earth might be withering under the 860°F (460°C) surface temperatures of Venus or frozen by the minus 350°F (−212°C) conditions found on Neptune.

Composition

As it turned out, the sun became a star of relatively modest size and heat, creating conditions on Earth temperate enough to let the oceans form while delivering a steady supply of about 1,400 watts of energy to each square yard of the planet. The sun's roughly 10,000°F (5500°C) surface gives it a yellowish hue and makes it a G2-type star, near the middle of the stellar classification scheme for color and temperature though a bit brighter and hotter than most other stars in the Milky Way. It is considered a yellow dwarf, lying amid what astronomers call the main sequence of stellar life cycles.

Like all stars, the sun is primarily made of hydrogen, the fuel for the atomic reactions at its core. Hydrogen represents just over 92 percent of the sun's matter, with helium making up nearly 8 percent, and trace amounts of sodium, iron, and other elements accounting for the small remainder. What we see from Earth—the outermost visible layer known as the photosphere—has substantially more helium, about 25 percent, with about 74 percent hydrogen and small amounts of oxygen, carbon, iron, sulfur, and many other elements.

Protagonist of the Sky

The sun is unremarkable in the context of other stellar objects, yet it has fascinated humans for centuries. It was treated as a powerful god by the ancient Egyptians and Aztecs, and it figures as a deity or personified figure in cultures worldwide. The sun is weaved so deeply into our daily lives that it not only governs external conditions like the weather but also triggers our internal synthesis of vitamin D and can sway human psychology.

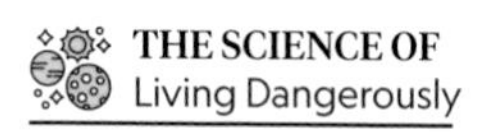

Living, as Earth does, in even rough balance with a star is inherently treacherous. The sun's radiation output contains high-energy, life-damaging gamma rays, x-rays, and ultraviolet rays. Much of the more dangerous stuff is buffered by Earth's atmosphere and magnetic field. Still, depletion of the planet's upper ozone layer is linked to rising skin cancer rates as more ultraviolet light reaches Earth's surface. Warnings about sun exposure and the need for protective clothing or sunscreen have increased in proportion.

THE LIFE OF THE SUN

AVERAGE IN ITS STRUCTURE compared with other stars, the sun has an average life expectancy too—about 11 billion years, which puts our solar system squarely in its middle age. A smaller, cooler-burning star, such as a red dwarf, would live for much longer, perhaps into the tens of billions of years, whereas a hot supergiant star might spend its fuel in only a million years or so.

Our star probably has about five billion years of primary fuel remaining, and another few hundred million years before it fades from the sky. That is still an unimaginable amount of material. Each second, nuclear fusion in the sun consumes about 700 million tons of hydrogen, creating 695 million tons of helium in the process. The excess five million tons of matter are converted directly into energy. But, one day, the tank will run dry. What happens then?

FURTHER

The sun accounts for about 99.8 percent of the solar system's mass and, like the rest of the universe, is made almost entirely of hydrogen and helium. It does contain trace amounts of other elements, including iron, nickel, oxygen, silicon, sulfur, magnesium, carbon, neon, calcium, and chromium.

Beginning of the End

The endgame of a star's life depends on how it lived—primarily on its size and temperature. In the case of our yellow dwarf sun, as it begins to run out of its chemical fuel, the amount of energy pouring outward will begin to decline.

An imagined view of Earth's barren surface after the sun expands to a red giant

The equilibrium established billions of years ago between outward-moving energy and the inward, unrelenting crush of gravity will be upset. As gravity continues to assert itself, the sun will begin to collapse.

That contraction will raise the sun's inner temperature and give the star a last gasp of brilliance. Leftover hydrogen, in the zone just outside of the sun's fusion-furnace core, will begin to burn, and the helium created from billions of years of nuclear reaction will also ignite, fusing into carbon.

Death of the Sun

The energy from these new reactions will cause the sun to expand far beyond its current boundaries, turning into a red giant. This engorgement will lead to the sun's engulfing an area likely to include the current orbits of Mercury and Venus—and perhaps swallowing both planets along the way. By this point, the sun will have boiled away all of Earth's oceans, as well as its inhabitants. The planet itself, as well as others in the solar system, will move outward into farther orbit as the sun spews its gas and matter into space and its gravitational hold continues to weaken. The star might begin to pulse, turning into a variable star. In the next phase, ejected gases will form a cloud—a nebula—as they begin to dissipate.

When the second round of fuel consumption comes to an end, the fusion reaction will halt altogether and gravity will again begin pulling the sun's matter into itself. The remaining material, including whatever is left of the naked hot core, will be compressed into an area about the current size of Earth. It will then become a white dwarf—a drifting, cooling remnant of a once-unique corner of the universe.

THE SCIENCE OF
Nuclear Alchemy

Nuclear power plants get their power from fission—splitting atoms to release energy from the bonds that hold subatomic particles together. The sun relies on fusion, a powerful process in which atoms of one element combine under heat and pressure to form atoms of a new element, releasing energy in the process. With stellar fusion, four hydrogen nuclei combine into a single atom of helium. Measured by mass, the helium atom is smaller than the four original hydrogen atoms by about 0.7 percent. Converted directly to energy, that proportionately small amount of leftover matter powers our universe.

ANATOMY OF THE SUN

FROM OUR VANTAGE POINT the sun seems placid. Though its heat can be intense, the view from Earth is of an object that shines as it moves through the sky. But ever-more-sophisticated spaceborne probes and scientific instruments reveal that it's a roiling, complicated place of violent extremes, with its own weather systems and an almost disturbing unpredictability.

The whole star rotates but in an eccentric way: Its equatorial region moves as if it is a solid, completing a turn in 24 Earth days, whereas the polar regions lag behind, taking 30 days or so for one rotation. Along with the regular flow of energy, phenomena like solar flares and coronal mass ejections send billions of tons of atomic material hurtling toward Earth—enough to affect our communications and electrical systems—and, potentially, our health.

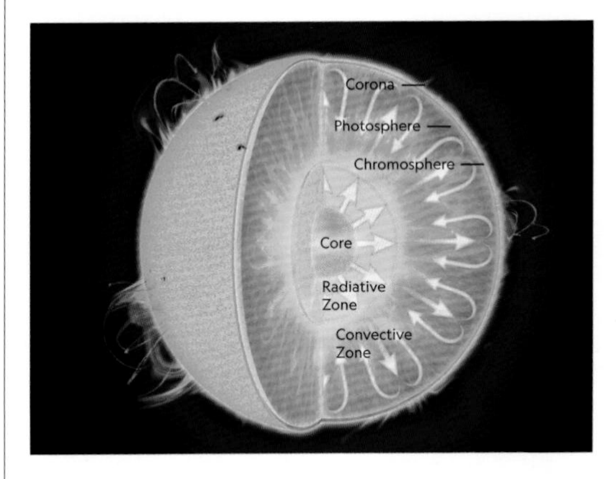

Inside Our Broiling Star

The heart of the sun is its dense, high-pressure core, where roughly half of the star's mass has been squeezed into about 7 percent of its volume. It is here that the process of nuclear fusion takes place. Temperatures can reach an astounding 27,000,000°F (15,000,000°C). As hydrogen is converted to helium, packets of energy called photons are emitted and begin a journey to the surface that can take millions of years to complete.

These photons fight their way from the core, beginning to lose energy and cooling as they travel through a radi-

FURTHER

Nuclear fusion of hydrogen into helium in the sun's core generates about 400 trillion watts of energy each second.

THE SCIENCE OF NASA Parker Solar Probe

Parker Solar Probe has been designed to travel closer to the sun than any spacecraft before it, flying as close as 4 million miles (6.4 million km) above the sun's surface. Facing unprecedented heat and radiation like no spacecraft before it, the compact car-size probe will provide us with a better understanding of the solar corona and solar wind, which will increase the accuracy of our ability to forecast major space weather events that impact life on Earth. The mission has 24 flybys planned that will run through the middle of 2025.

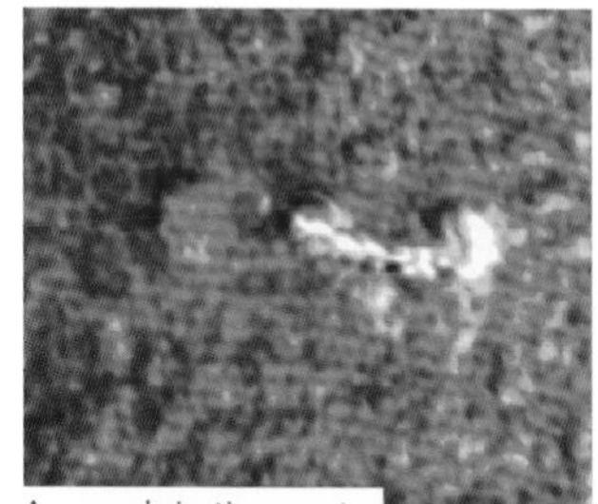
A coronal ejection erupts.

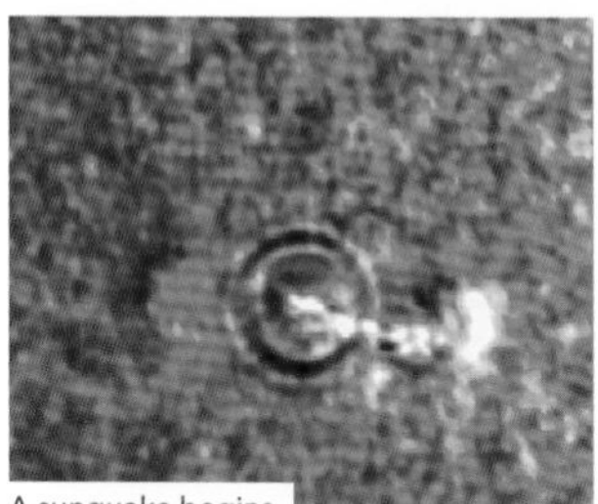
A sunquake begins.

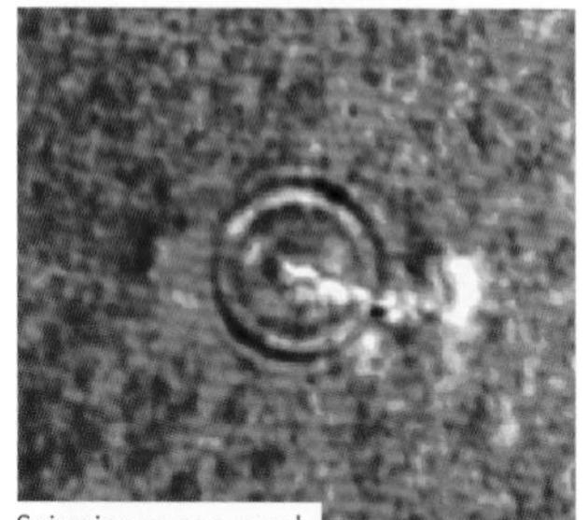
Seismic waves spread.

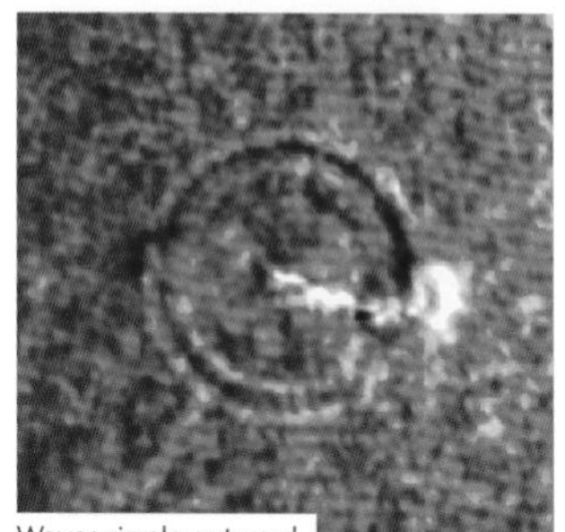
Waves ripple outward.

ated zone. About three-fourths of the way to the surface they pass an area where convection currents continue drawing gas and heat outward. They then reach the photosphere—a roughly 300-mile-thick (500 km) layer. At this point, temperatures have fallen to around 11,000°F (6093°C).

The brilliant, steady light emitted at the photosphere makes the sun appear solid—not as a turbulent and multilayered surface. But photos of the sun reveal a grainy surface that appears to be in rapid boil, with bubbles of gas pushing to the surface, settling out and sinking back into the stew over about 10 minutes' time. A pinkish layer called the chromosphere surrounds the sun's outer surface, firing out spicules, or spikes, of gas. It is visible only when the sun is in total eclipse (and when special equipment is used to protect the naked eye). Solar prominences—clouds of gas—float across the surface or shoot out in arcing geysers.

Surrounding that, the sun's far-reaching corona drifts for millions of miles—a ghostly and, except during total solar eclipse, invisible halo of gas. In the corona, temperatures inexplicably rise back to perhaps 2,000,000°F (1,000,000°C)—this increase remains one of the biggest mysteries in astronomy. Compared with the dense solar core, the corona is almost nonexistent—trillions of times less dense than the air on Earth. Holes in the corona, caused by the sun's magnetic field, are where steady streams of particles known as solar wind break free and flow toward the rest of the solar system.

SUNSPOTS & SOLAR FLARES

IN THE MID-19TH CENTURY, amateur astronomer Richard Carrington determined that patchy spots near the sun's equator rotated around more quickly than did spots located farther from the middle. In 1859, he also noted—as did amateur sun-watcher Richard Hodgson—the eruption of a patch of white light near a sunspot group, an event that coincided with telegraph system problems as well as a brilliant display of auroras. The significance of those observations was not clear at the time. In fact, the phenomena involved is still being studied by scientists trying to determine the interplay between the sun's weather and its magnetic field.

The Sun's Rotation

Carrington's observations were explored more fully by the European Space Agency's Solar and Heliospheric Observatory (SOHO). In the mid-1990s, data from SOHO showed that the sun's dense but still gaseous center rotated like a solid, while the outer areas, particularly those near the sun's poles, rotated much more slowly. That differential rotation stretches the sun's magnetic field lines, which get tangled even further by the continuing surge of gas through its convection zone. Those disrupted currents can slow the flow of solar material, causing it to cool and darken.

FURTHER

Astronomers have recorded the appearance of sunspots as far back as 28 B.C., when Chinese astronomers made note of them.

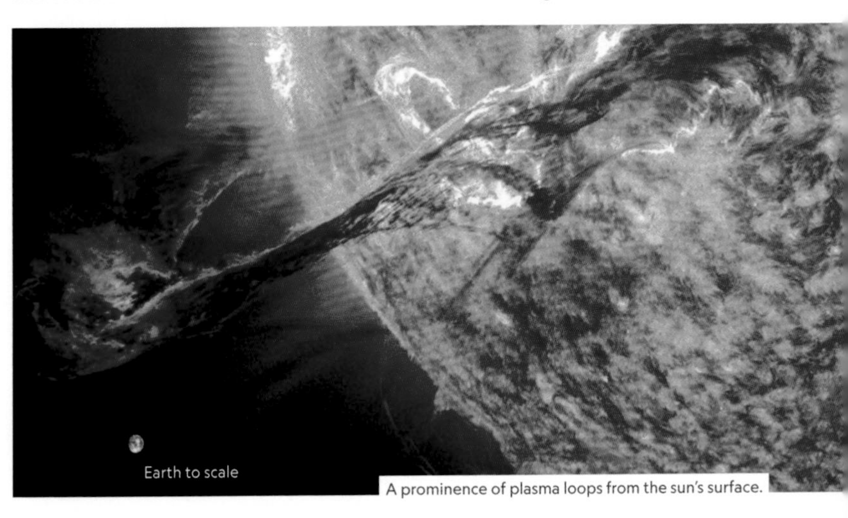

A prominence of plasma loops from the sun's surface.

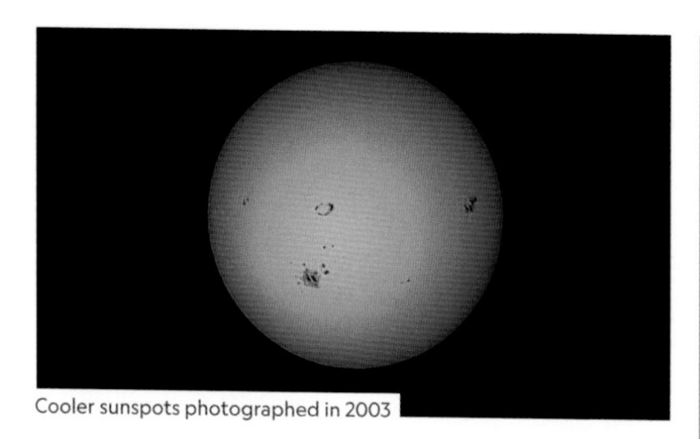
Cooler sunspots photographed in 2003

Appearing as a dark splotch on the surface, sunspots can be as much as 3600°F (2000°C) cooler than the sun's typical surface temperatures of around 10,000°F (5500°C).

Brilliant Rays

Typically clustered near the sun's equator, sunspots peak and ebb in an 11-year cycle that coincides with other solar activity. The buildup of magnetic tension eventually gives way, and its release can eject billions of tons of atomic particles in a solar flare. This barrage of charged particles can knock out communications equipment and satellites—and disrupt the flow of electricity in terrestrial power grids. It is also what causes the brilliant auroras that can be seen on Earth (see page 40).

Because of the implications for earthbound technology and systems, predicting solar storms has become a priority for agencies like the National Aeronautics and Space Administration (NASA). In 2010, NASA launched the Solar Dynamics Observatory (SDO), which connects information about the movement of charged plasma in the sun's interior with changes in the magnetic field closer to the surface—and ultimately will try to develop models to predict when solar storms are brewing.

The study of the sun's magnetic field extends beyond what's going on with the solar surface. Indeed, though the sun's magnetic field is considered weak overall, its influence is felt far into space. The so-called heliosphere extends perhaps 100 times the distance between Earth and the sun (100 AU), providing a sort of curtain between our solar system and the incoming interstellar wind.

THE STORY OF the Little Ice Age

Around 1645, something strange happened: Sunspot activity came to a virtual halt and remained almost dormant for about 70 years. This sunspot drought is known as the Maunder Minimum, and it coincided with the coldest period of the Little Ice Age that settled in across the Northern Hemisphere during that time. Associated with advancing glaciers, lost farmland, and other disruptions, the epoch showed how even a small change in the sun's activity can affect our planet. It is estimated that solar output during that period dropped by just a quarter of one percent.

THE SUN'S PATH

Time-lapse photography reveals the sun's arc.

POSITIONS OF THE SUN

SPRING EQUINOX
DATE: 03/20 OR 03/21
SUN'S POSITION: Equator (0°)

SUMMER SOLSTICE
DATE: 06/20 OR 06/21
SUN'S POSITION: Tropic of Cancer (23° 27′ N)

AUTUMN EQUINOX
DATE: 09/22 OR 09/23
SUN'S POSITION: Equator (0°)

WINTER SOLSTICE
DATE: 12/21 OR 12/22
SUN'S POSITION: Tropic of Capricorn (23° 27′ S)

THE SUN'S APPARENT JOURNEY through the sky—and the changes to its route over the course of a year—provided an early sense of how Earth's position in the universe affects daily life. Of course, it took centuries to sort out that Earth is the planet doing the moving. Still, the daily change in the relative position of the two bodies has long found practical applications, providing a way to tell time, judge the seasons, and mark annual celebrations.

The sun's path through the sky, known as the ecliptic (the word also refers to Earth's route around the sun), was decided by the origins of our solar system. As the sun's birth cloud began to spin and take shape, it flattened into a disk. With hydrogen settling at the core, other material collected farther out and coalesced into what would become the planets. These planetary bodies would soon orbit in the same direction as the original spinning solar cloud, all aligning on roughly the same plane.

Earth, as we know, moves in two distinct ways—it rotates on its axis while also progressing steadily on its annual

THE STORY OF
the Pueblo Sun Temple

Ancient peoples in the southwestern United States developed systems for tracking events like solstices and equinoxes. A ceremonial Pueblo Sun Temple, in what is today Mesa Verde National Park, features slits in the walls that could be used to precisely locate positions on the horizon. Two alignments are within one degree of the rise of Pleiades around A.D. 1250, when the temple was in use, a remarkable design to execute without written language or a numeral system.

orbit around the sun. The planet's daily spin moves in what from our vantage point is west to east. That makes the sun and stars appear to move in the opposite direction, from east to west, across our sky.

Changing Seasons

The orbit around the sun, meanwhile, is marked by Earth's axis of rotation, which is tilted about 23.5 degrees off perpendicular from the ecliptic. That tilt is what accounts for the seasons and for the lengths of time the sun appears in the sky throughout the year.

On the first day of summer in the Northern Hemisphere, typically June 21, the northern half of the planet is tipped toward the sun, which is why the North Pole experiences 24-hour daylight. Night still occurs farther south, but the day is the longest of the year, with the sun at its highest point in the sky. Over the course of six months the sun's path will flatten, the days will shorten, and the whole hemisphere will transition from summer into autumn. Below the Equator, the same dynamic is at work—but the dates are reversed. In the Southern Hemisphere June 21 marks the shortest day of the year, as that part of Earth is tipped away from the sun.

Equatorial areas, of course, experience none of these extremes. Located around Earth's middle, they have days and nights that stay roughly the same throughout the year, with the seasons offering little variation.

Seasons occur on all the planets. Both Venus and Jupiter tilt only three degrees so there is minimal difference between seasons. Mars has an elliptical orbit that draws the sun nearer then farther away, causing a much stronger effect than seasonal changes on Earth.

FURTHER

Every summer solstice many revelers make a pilgrimage to Stonehenge in southwestern England. Considered one of the must-see architectural wonders of the ancient world and dating back some 4,500 years, the stone circle is made of 6-foot (2 m), 4-ton (3.6 t) ancient bluestones, placed in such a way that they align to frame the sun's rays as it rises on the longest day of the year. Some scholars assume that this megalithic monument may have also been used as an astronomical calculator to predict the path of the sun, the moon, and perhaps even their eclipses.

VIEWING THE SUN

WALK OUTSIDE on a sunny day and the natural reaction is to shield your eyes from the intense light. It is a well-founded protective instinct: A direct view of the sun can damage the naked eye. Channel that light through a telescope or binoculars and the effect can be quick and catastrophic: blindness. It is unfortunate that the closest star is also the one celestial object that requires safety precautions to view in detail. But the point needs emphasizing, particularly where kids are involved. Caution should be the watchword, lest anyone be tempted to flash their binoculars skyward or move under the telescope to "take a peek."

Eclipse-watchers use protective solar filters.

Safe Observation

Some simple rules and a bit of extra equipment will let you view the sun safely—watching its granular surface bubble, tracking sunspots, and perhaps catching a glimpse of an erupting solar flare. These are events that can be seen during the day—the earlier the better, preferably from a grassy area or across a body of water, where the air will remain cooler and more settled and there is less light pollution from cities.

If you are not using binoculars or a telescope, safe solar viewing can be done through specially made solar viewing

FURTHER

For backyard telescope users, the sun will veritably come alive when viewed through the crimson light of a hydrogen alpha filter. Unlike normal white light filters, which show off our stars' photosphere and its sunspots, H-alpha filters allow observers to get right where all the action is: the chromosphere. By filtering out everything but the narrow sliver of light emitted by hydrogen, sun-watchers can see things like wispy prominences and fiery flares.

glasses or a plate of number 14 welder's glass. These glasses may not be on the shelves of major retail outlets but can be easily procured through a specialty astronomy supplier. With this protection, you can hunt for sunspots with the naked eye.

Through a Telescope

If you don't expect to do regular solar viewing, the cheapest way to stay safe with your telescope is to use it (or tripod-mounted binoculars) as a makeshift projector. To orient your equipment, use its shadow as a reference point, adjusting until the shadow is at its smallest. Again, do not look through the equipment for that initial orientation—and if your telescope has a finder attached, make sure to cover the aperture so no one can look through it. Using a stand or clip, affix a sheet of white paper or cardboard a foot or so behind the eyepiece. You'll see the sun's projection and can focus to sharpen the view. Cut a hole in the middle of a piece of cardboard to fix onto the barrel of the scope in a way that shields the "screen" from ambient light.

When using this method, remember to protect your equipment. Heat from the sun can build up inside the telescope tube and damage it. A telescope aperture should be covered with cardboard, or other shield, with a hole that reduces the open area to no more than 4 inches (100 mm)—going as small as 2 inches (50 mm) isn't being too cautious.

Filters

For more serious viewing, the easiest solution is to invest in a solar filter: a piece of metal-coated glass or plastic that will cut the sun's brightness by a factor of perhaps 100,000. (Never use a solar filter that only fits over the eyepiece; these crack without warning under the solar heat, causing serious harm to the observer's eye.) Less expensive filters are made of aluminum-coated Mylar, available through specialty astronomy retailers, and can even be hand-cut from sheets of the stuff. Glass filters will provide more realistic colors and can be bought in sizes to fit the aperture of your telescope.

SKY-WATCHERS: Solar Blindness?

Galileo Galilei spent a lot of time behind a telescope in the early 1600s, deducing some of the basic facts about the solar system. He was also an avid sunspot counter in the days well before Mylar plastic and H-alpha filters. Did he go blind? Probably not. Well, not until he was in his 70s, and the condition probably stemmed from illness or a chronic condition rather than from damage caused by the sun. He most likely used the projection method to avoid direct observation. But there are famous cases of sun damage among astronomers, including Sir Isaac Newton, who for several months was troubled by a "phantasm of light and colors" caused by looking at the mirrored reflection of the sun in a dark room. Such exposure can cause a scotoma—a temporary blind spot that should fade with time.

| ALL ABOUT ECLIPSES

AN ECLIPSE IS A BREATHTAKING and spooky sight. The darkness-at-noon experience of a total solar eclipse, in particular, is well worth seeking out. But even the more common lunar eclipse is a stunning sight and a great showcase of celestial mechanics.

What Are Eclipses?

An eclipse happens when either the moon or Earth blocks the light from the sun. In a lunar eclipse, Earth lies directly between the sun and the moon so that Earth's shadow falls on the moon. This, of course, can happen only during a full moon, when it is on the opposite side of Earth from the sun. In a solar eclipse, the roles are reversed: The moon comes between the sun and Earth, and the shadow of the moon falls on Earth. This can happen only during a new moon.

You'd think that each kind of eclipse would happen once a month, but in fact each occurs about twice a year. This is because the sun, moon, and Earth do not orbit in exactly the same plane. The moon's orbit is tilted about five degrees from that of Earth's, so eclipses are possible only when the plane of its orbit intersects the plane of our orbit. In any year,

The sun's corona revealed during a solar eclipse

A full moon at the beginning of a lunar eclipse

there are between two and seven eclipses, solar and lunar combined.

When the sun shines on the moon or Earth, the resulting shadow cone has two parts: a narrowing inner "umbra," where light is completely blocked, and a wider outer "penumbra," where light is only partially blocked. Total eclipses occur within the umbra, and partial eclipses in the penumbra. During a lunar eclipse, Earth's shadow envelops the moon and the darkened moon can be seen from anywhere on the nighttime side of our planet (if the sky is clear). But during a solar eclipse, the moon casts a relatively small shadow on the much bigger Earth, so a total solar eclipse can be seen only along a small band on the turning Earth, an area that changes from eclipse to eclipse. It's important to know your location when planning to view an eclipse for this very reason.

A solar eclipse is a marvelous thing to see, but never look at the sun directly or you risk serious eye damage (sunglasses don't provide enough protection). View the eclipse indirectly, through projection, or using approved solar or number 14 welder's glass.

SOLAR HIDE-AND-SEEK

SOLAR ECLIPSES COME in three varieties: annular, partial, and total. Annular eclipses occur when the moon is too far from Earth to completely cover the sun; its inner shadow, the umbra, falls short of Earth's surface. During an annular eclipse, the moon's disk cuts into the sun and eventually crosses its face, but the sun's bright edge is visible all around it, forming a ring, or annulus (hence the name). Your surroundings will grow dim, but not dark.

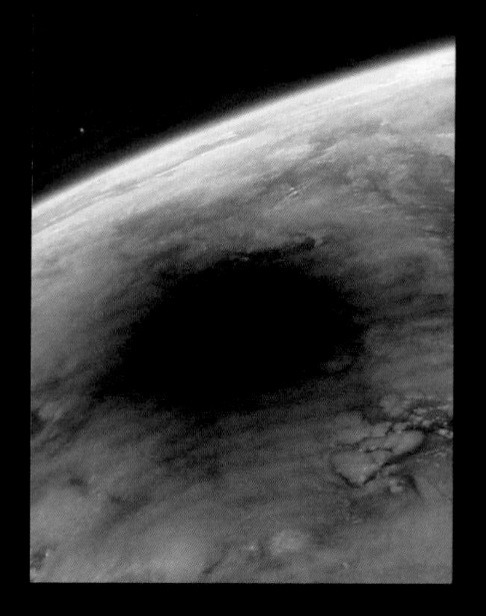

How a Solar Eclipse Works

Partial solar eclipses take place when your viewing site lies within the moon's outer shadow, its penumbra. They are typically undramatic, dimming the sunlight only a

When the moon passes between Earth and the sun, the shadow cast on Earth (above) causes a solar eclipse.

bit, but with proper eye protection you can see a bite taken out of the sun.

The showstopper of eclipses is the total solar eclipse. These are visible only within the moon's umbra, when the moon's disk appears to completely cover the sun. The path traced by this dark inner shadow, known as the path of totality, is typically about 200 miles (300 km) wide and often narrower. The slender trail of a total solar eclipse varies from year to year in an 18-year pattern. In August 2008, for instance, it crossed the Arctic, Greenland, and Russia, and in 2018 it curved diagonally across the United States from Oregon to South Carolina. The next total eclipse to cross North America will take place on April 8, 2024. The path of totality will take a more easterly route, starting in Mexico before entering the United States by way of Texas, then traveling up the northeast and crossing into Ontario and Quebec, Canada. See the back inside cover for the schedule of solar eclipses in North America.

A total solar eclipse is an extraordinary experience. During totality—when you are in the central shadow of the moon—daylight fades to dark twilight. Birds roost, crickets chirp, and the brighter stars and planets appear in the sky during those precious few moments of darkness. Sky-watchers can see the sun's spectacular corona, its fiery outer atmosphere, feathering out around the edge of the moon's disk. It is safe to look at the eclipse with the naked eye only during totality, but if there's any uneclipsed sun visible, you must use a proper solar filter.

Explanations From Mythology

Many cultures have understandably feared solar eclipses and believed they were signs of bad fortune. Solar eclipses have figured in tales as diverse as the Chinese legend of the devouring dog or dragon that took bites from the sun (local residents would bang pots or shoot arrows to scare the beast away). In other cultures, a frog, a bear, or even a werewolf was blamed for eating the sun. The sun and the moon are often interpreted as a feuding brother and sister or a married couple striving to make peace. In this instance, the Batammaliba people of Togo and Benin offer a solution: The solar eclipse is a call to resolve old feuds and settle disagreements to encourage the sun and moon to stop fighting.

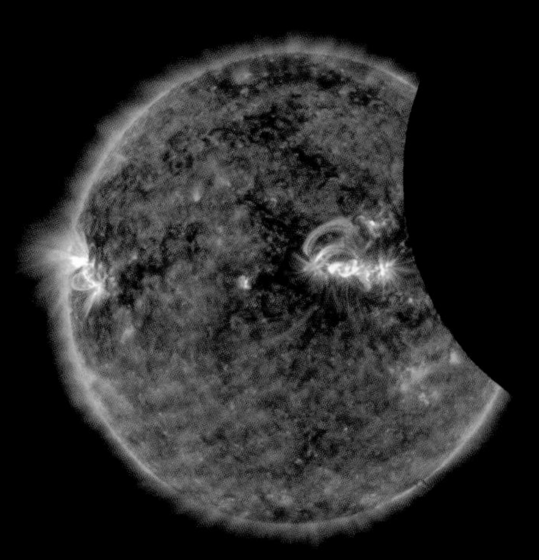

A partial solar eclipse captured by a spacecraft

LUNAR DISAPPEARING ACT

LIKE THE SUN, the moon experiences three kinds of eclipses. If it passes through the penumbra—the outer region of Earth's shadow—we see a penumbral lunar eclipse. It can be hard to detect with the unaided eye because the moon will typically be only a bit dimmer than usual.

If just some of the moon passes into the umbra, or partially in the penumbral as well, it undergoes a partial lunar eclipse. In this case, Earth's dark inner shadow carves a passing bite out of the moon's disk.

When the moon moves completely into Earth's umbra, we see a total lunar eclipse. Even then the moon will not become completely dark, like the sun does during a total solar eclipse. The full circle will be visible but the brightness will appear much dimmer.

Total lunar eclipses can last as long as an hour and a half or pass by in a few minutes. You can view them safely with the naked eye. Binoculars are helpful; with a telescope, use a low-power eyepiece so you can see the whole moon at once.

Blood Moon

During the total phase of the eclipse, while most sunlight is blocked from reaching the moon, some stray sunlight travels through Earth's thin ring of atmosphere and gets projected onto the moon's surface. As this sunlight moves through the column of air, however, it

Earth casts a rust-tinted shadow on the moon.

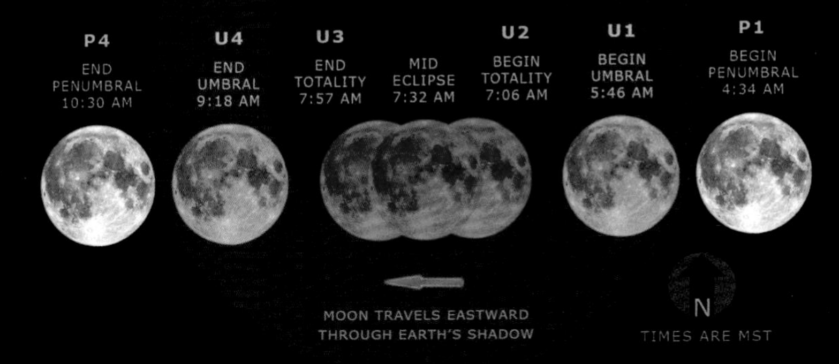

The stages of a lunar eclipse

gets bent, or refracted, by dust and pollutants, shifting the sunlight toward the red part of the spectrum. It's the same effect that creates the typical colors of sunsets and sunrises we see around the world. As a result, the sunlight that leaves our atmosphere and hits the moon paints the darkened lunar disk an orange to rust-red hue. Without the atmosphere, a lunar eclipse would be completely black. If Earth's atmosphere is unusually dusty, blocking more sunlight, the fully eclipsed moon may grow darker red, hence the popular name "blood moon." At other times, when Earth's upper atmosphere is free of particulate matter, it may stay fairly bright and orange. Active volcanoes spewing tons of ash into the upper atmosphere have historically been known to cause blood-red eclipses.

Eclipse Omens

Sky-watchers have tracked lunar eclipses since ancient times; both the Chinese and the Maya, for instance, kept careful tables of such occurrences. In 1504 Christopher Columbus, using his astronomical tables, famously tricked the inhabitants of Jamaica. Stranded on the island and needing food from the native Arawak (who were tired of their mistreatment by the European sailors), Columbus consulted his almanacs and noted that a total lunar eclipse was due. He told the Arawak that God was displeased with them and would darken the moon with his wrath. Sure enough, the moon turned the characteristic bloody red of a lunar eclipse. The locals promised complete aid if only God would relent, which he appeared to do about an hour later.

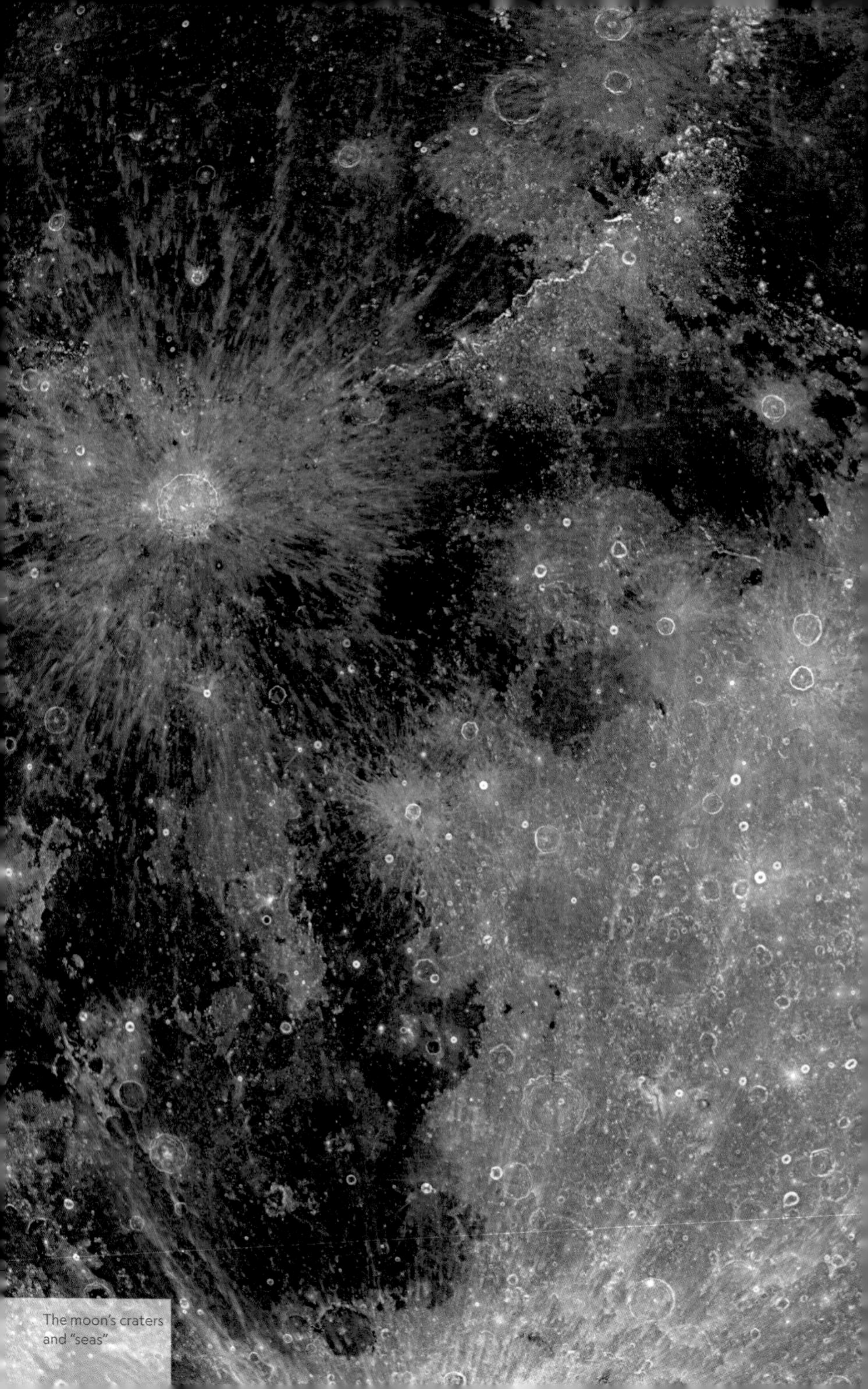

The moon's craters and "seas"

CHAPTER 4

THE MOON

OUR NATURAL SATELLITE

SYMBOL: ☾

RADIUS: 1,079.6 mi (1,737.4 km)

MASS: 7.3483 × 10^{22} kg

DISTANCE FROM EARTH: 238,855 mi (384,400 km)

LENGTH OF YEAR: 29.5 days

LENGTH OF DAY: 27.3 days

ORBITAL CIRCUMFERENCE: 1,499,619 mi (2,413,402 km)

THE MOON IS A GREAT FRIEND to stargazers, outshining even the worst light pollution. While it can easily be viewed with the naked eye, binoculars will bring out many of the moon's details and amazing features.

The moon orbits at an average distance of about 239,000 miles (384,600 km) from us. Its 2,159-mile (3,475 km) diameter is roughly a quarter of Earth's, but because it is composed of lighter elements, it has about an eightieth of our planet's mass and one-sixth of its gravity.

Lunar Origins

Evidence from lunar missions suggests that the moon was formed when a Mars-size object struck young Earth, creating debris that eventually congealed into the pale orb we know today. That theory accounts for several things

A cosmic collision billions of years ago ejected Earthly debris that became the moon.

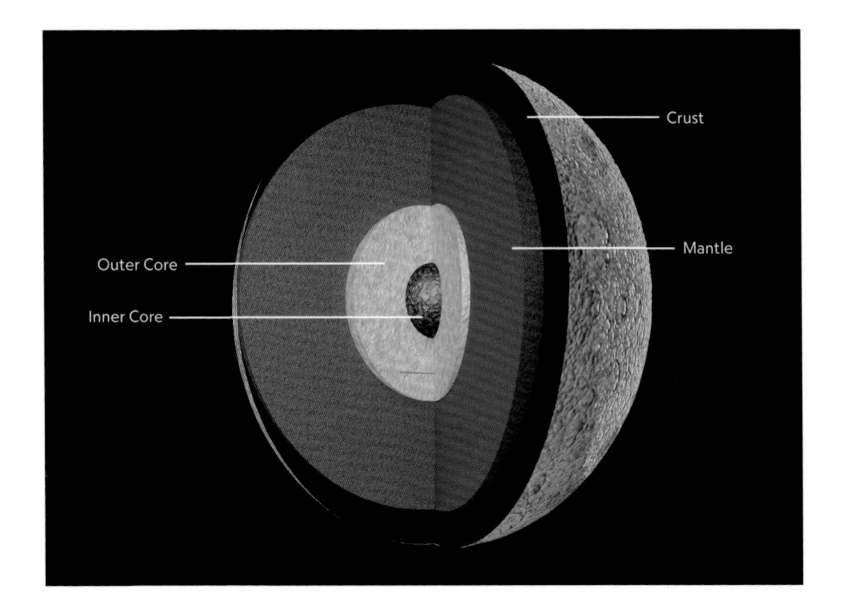

about it: the presence of rocks similar to those found near the surface of Earth; the lack of water (vaporized during the explosion, though trace amounts have been found in lunar rocks); and aspects of its orbit.

Lunar Desolation

Like Earth, the moon is made up of a core, mantle, and crust. Its core is made of almost solid iron, encased in a thin shell of semi-molten material and, beyond it, a layer of less dense rock. Its surface is a perfect map of its history, with no weather to change its craters and mountains. Meteor collisions over its 4.5-billion-year life span have created a surface of loose and pulverized rock, called regolith. Its craters span as wide as 1,600 miles (2,575 km) across, with walls as high as 4.8 miles (7.7 km). Its distinctive "seas"—so called because early observers thought they might be the beds of dried-up oceans—were actually created by molten lava brought to the surface by major impacts that occurred around 3.8 billion years ago. Despite its small size, the moon's tectonic and volcanic activity has shaped its impressive landscape into vistas that are both similar and alien to those found on Earth.

FURTHER

The moon's power is always pulling at our planet. Its gravity tugs slightly on Earth's oceans when they face it, which is what creates high tides.

THE MOON IN MOTION

BECAUSE THE MOON always keeps the same side exposed to Earth, it can appear a somewhat passive traveler, trapped in its orbit and asserting little influence on its much larger neighbor. But the relationship between Earth and its satellite is an interesting and active one, each body influencing the other in a surprisingly complex partnership.

FURTHER

The Apollo 8 astronauts were the first people to see the far side of the moon with their own eyes. According to the NASA transcript, William Anders commented, "The backside looks like a sand pile my kids have been playing in for some time. It's all beat up, no definition, just a lot of bumps and holes."

The Moon's Rotations

Though it appears stationary, the moon is actually spinning on its axis, just as Earth does. Its rate of rotation matches the rate of its progress around Earth—both take about 27.3 Earth days to complete. This synchronicity occurred because, as the two bodies evolved, Earth's gravity created a bulging "land tide" in the moon, causing its spin to slow and synchronize with the speed of its orbit. For earthbound observers, that means only one side of the moon is ever visible—the near side—while the other side permanently faces away. It takes a bit longer, about 29.5 Earth days, to complete what is known as the moon's synodic month: the time needed to orbit Earth and return to the same position relative to the sun, completing a full lunar cycle.

Often referred to incorrectly as the "dark side," the hidden half of the moon isn't actually in constant darkness. It experiences phases just like the near side does, with a time each month when it is in full sunlight and one when it is in full shadow. The moon's far side was first observed and mapped in 1959, when Russia's Luna 3 spacecraft took pictures of it, followed by detailed maps created over many decades. This Russian feat is reflected in some of the names still attached to the moon's far side features, including the Moscow Sea and a crater named in honor of Russian cosmonaut Yuri Gagarin, the first person to orbit Earth. The far side looks very different from the near one, dominated by cratered highlands. Its significantly thicker crust prevented

The far side of the moon

Earth shadows the moon from the sun, causing moon phases.

magma from bubbling up when meteors hit it, which is why it doesn't have volcanic seas like the near side.

Mutual Orbits

It is a bit misleading to say the moon orbits Earth. Though the moon's gravity is only about one-sixth that of our planet, the two essentially orbit each other. Think of it as a spinning baton, albeit a lopsided one. Because Earth is so much larger and denser, the balancing point of the system—known as the barycenter—lies about 1,100 miles (1,770 km) below Earth's surface. This center of gravity for the Earth-moon system is what actually tracks the ecliptic plane around the sun, with the rest of the planet wobbling around it.

The moon's appreciable gravitational tug proved important in planning the manned Apollo missions. The flights followed a trajectory that placed them ahead of the moon, and they relied on its gravity to shoot them around to the far side. By reducing speed, the spacecrafts were able to get tugged along in lunar orbit. In the case of Apollo 13, the mechanics of trajectory were what saved the crew's life. After an oxygen tank exploded about 200,000 miles (320,000 km) from Earth, backup power systems boosted the craft into a free-return trajectory. Earth's gravity, and Kepler's laws of motion, helped it loop around the moon and toward home.

THE SCIENCE OF Moving Away

The moon is slowly spiraling away from our planet at a rate of about 1.5 inches (3.8 cm) each year. That's because the moon's gravity is continually pulling on Earth, raising tidal bulges in our oceans that cause tidal friction, which takes energy out of Earth and slows its spin. This loss in angular momentum causes the moon to simultaneously speed up and slip farther away from our planet. This lunar drift was suspected for centuries, but only confirmed when scientists timed how long it took to bounce laser beams off mirrors left on the moon by Soviet and American missions in the 1970s.

LUNAR MOVEMENTS

FURTHER

Have you noticed that the full moon rising or setting near the horizon looks bigger than when it's in the overhead sky? We know that this is an optical illusion because cameras show the moon as precisely the same size regardless of where it is in the sky. Some explain the convincing illusion by pointing out that on the horizon we can compare the lunar disk with familiar objects in the foreground like trees, hills, and houses. Others have claimed the moon may look larger to us because as a species we are closely attuned to sights on the horizon, which could pose a greater threat compared with those above (no flying lions on the savanna!).

JUST AS EARTH'S ORBIT changes the sun's apparent path in the sky, the moon's trip around Earth determines where it rises and sets, its course, and how much of its surface we can see. It takes just over 29 days for the moon to pass once around Earth, which translates into a daily shift of about 12 degrees. If you track the position of moonrise, you'll see it shift east in the sky by about that much each night.

Location in the Sky

The moon's orientation to the ecliptic also determines its height and path. The moon is offset from the ecliptic by around five degrees and thus holds close to that plane. In a Northern Hemisphere winter, the ecliptic is low in the daytime sky but arcs high and long through the night. The reverse is true in summer, when it tends to stay low in the night sky. During spring the ecliptic is angled sharply in the western sky in early evening, while in autumn the situation is reversed, with the plane angling upward in the east around dawn. These two times of year offer longer viewing times to spot and study waxing and waning crescents.

Phases of the Moon

The moon generates no light of its own; instead it functions like a projector screen, reflecting the light of the sun. As the moon travels, its relative angle to the sun and Earth

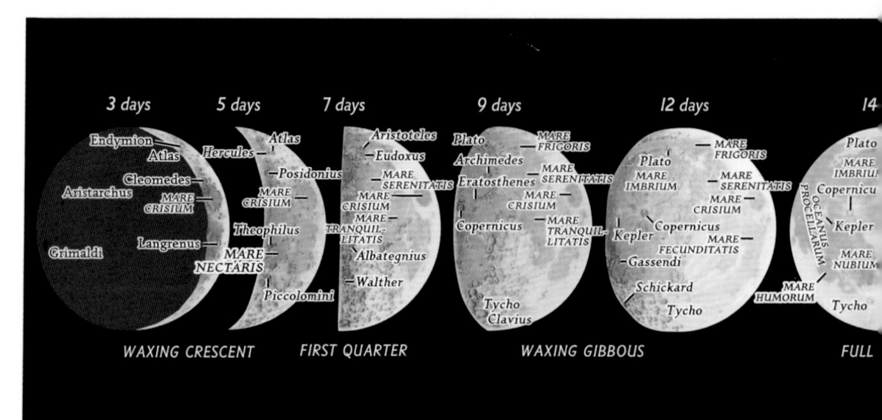

constantly changes, which is what creates its many different phases. At one extreme, the moon lies between the sun and Earth; fully backlit, the new moon can't be seen (unless the alignment produces a solar eclipse). At the other, the moon and sun sit on opposite sides of Earth, creating a bright full moon (or, if the alignment is right, a lunar eclipse).

The new moon moves gradually along its orbit, capturing a couple degrees of sunlight each night. Within a day or two, a thin waxing crescent appears. After about a week, the first quarter moon forms a roughly 90-degree angle with Earth and the sun, making it appear half lit—a phase called the first quarter moon because one-fourth of the lunar cycle has passed. Next comes a gibbous moon, from the Latin word for "hump," and then a full moon, which is sometimes bright enough to cast a shadow. After that, the lit portion of the moon begins to shrink, from a last quarter moon to a waning crescent, until it disappears into the dark and the next lunar cycle begins.

There is a direct connection between lunar phases and the moon's patterns of rising and setting. Because of the geometry involved, a full moon will rise with the setting sun, since they are directly opposite each other, and set only with the following day's sunrise. The moon has no seasons, since it runs nearly perpendicular to the ecliptic. Sunlight travels almost horizontally over its two poles.

THE STORY OF the Harvest Moon

Once a year in the Northern Hemisphere, sky-watchers can enjoy a phenomenon called the harvest moon. It happens around the autumn equinox when the moon's orbit makes a shallow angle to the horizon. That causes it to rise at about the same time several days in a row—around sunset—making it appear big, orange, and full. Moonrise comes around 50 minutes later each night throughout the year, but during this time that window shrinks down to about 30 minutes. Before electricity, this early rising moon was crucial because of the extra light it gave farmers as they harvested crops.

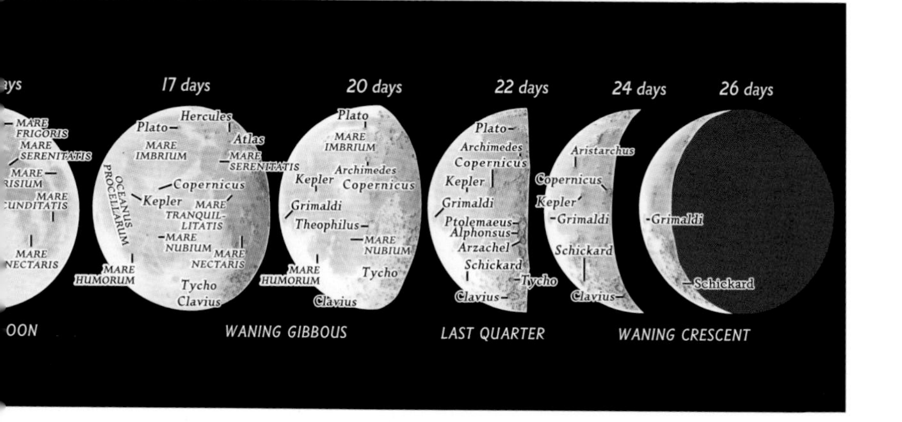

FACE OF THE MOON

MANNED MOON LANDING MISSIONS

APOLLO 11
07/16/1969
NOTABLE EVENT:
First manned lunar landing

APOLLO 12
11/14/1969
NOTABLE EVENT:
Exploration of Ocean of Storms

APOLLO 14
01/31/1971
NOTABLE EVENT:
Photography and exploration

APOLLO 15
07/26/1971
NOTABLE EVENT:
First use of lunar rover

APOLLO 16
04/16/1972
NOTABLE EVENT:
Surveying and sampling of Descartes region

APOLLO 17
12/07/1972
NOTABLE EVENT: Last lunar flight. Surveyed Taurus-Littrow region

AS WE BEGAN TO EXPLORE IT, the moon transformed from a little-known source of fascination to an object with an interesting, if convoluted, history. Thanks to the 12 people who have walked on its surface and an armada of robotic spacecraft, we know that history has been rather a violent one. Particularly in the first half-billion years or so, when debris left over from its formation regularly bombarded the moon's surface and created the lunar landscape we see today. Two types of terrain dominate the near side: 83 percent of it is covered by primordial, bright highlands, while the remaining 17 percent is covered by much younger, darker plains.

Smooth Seas

In the 1600s Galileo observed many of the moon's ridges and craters through a telescope, including dark areas that looked like calm, flat oceans. Ancient astronomers called them maria (pronounced MAH-ree-uh), the Latin word for "seas." Many Latin names still endure today on the moon's surface, the most famous being Mare Tranquillitatis, or the Sea of Tranquillity, where Neil Armstrong took his first giant steps. With binoculars, you can still observe the general region where his capsule, Apollo 11, landed in 1969. The vast majority of maria are concentrated on the side of the moon facing Earth.

These smooth maria offer a clue to understanding the moon's turbulent early history. It suffered repeated impacts

Moonscape of the Taurus-Littrow valley as captured by an Apollo 17 astronaut

from debris left over from its formation, producing holes that were hundreds of miles wide but relatively shallow at only 10 miles (16 km) deep. These impacts remelted some of the moon's remnant magma, filling the holes with smooth-looking molten rock. The frequency and size of these impacts have decreased markedly over time as debris has cleared from interplanetary space, leaving these areas mostly intact and very valuable as a historical road map.

Viewing Craters

Some of the collisions simply punched holes in the moon's surface, pocking it all over with craters. Scientists estimate that the upper few miles of crust, which appear light in color, were pulverized and reshuffled multiple times in the moon's early history. The last major crash has been dated to about 109 million years ago, which created the crater now called Tycho, after astronomer Tycho Brahe. Individual craters can be dozens of miles wide and several miles deep. To find the most recent ones, look for the ray-like sprays of rock that sometimes emanate from their base—evidence of the impact that caused them.

The moon's first and third quarter phases, when sunlight strikes it at an angle, bring out these details in sharp relief. You can see and study prominent craters, such as Copernicus and Theophilus, with your binoculars. Through a telescope, maria make excellent reference points to help you locate craters and other objects.

THE SCIENCE OF Lava Tubes

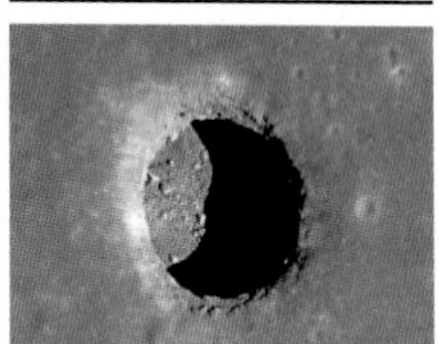

Scientists analyzing high-resolution photos from lunar orbiters have discovered lava tubes on the surface of the moon. Once filled with flowing lava, the channels eventually drained, leaving behind hollow tubes, and some have revealed their presence via "skylights": holes where the roof has collapsed. Some of them are the size of a city and could offer future lunar colonists secure living spaces and protection from harmful radiation, extreme temperatures, and meteorite showers. They could even offer access to subsurface ice—and with it, water.

THE MOON'S NEAR SIDE

THE MOON'S SYNCHRONOUS ORBIT with Earth—spinning on its axis in the same time it takes to move around the planet—means that only one side is ever visible to us. This visible side, or near side, features light-colored highlands where collisions with meteors and comets have left broad, deep craters. The large, dark maria are where molten rock has oozed out and filled large impact basins, leaving behind a smoother surface.

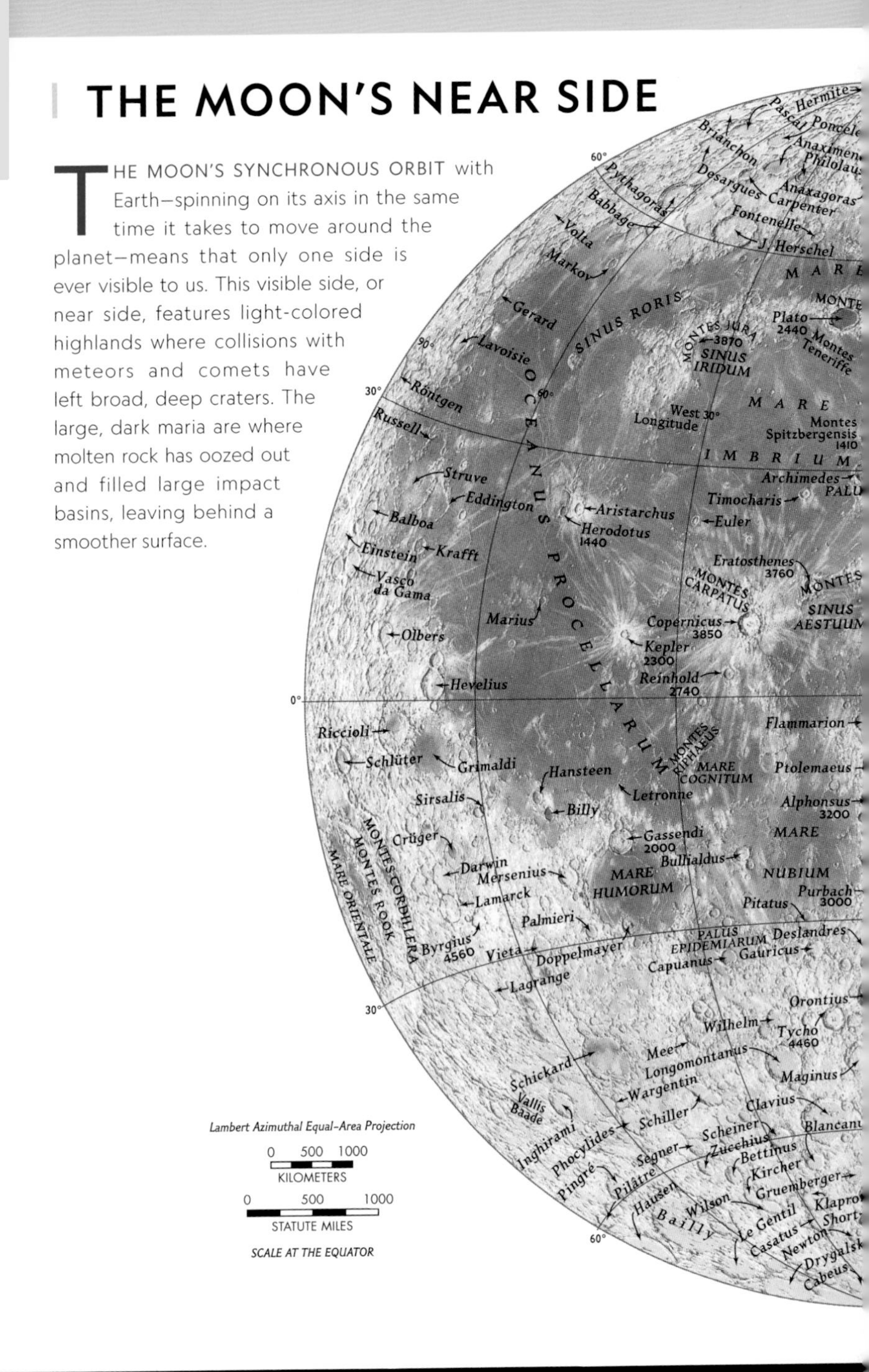

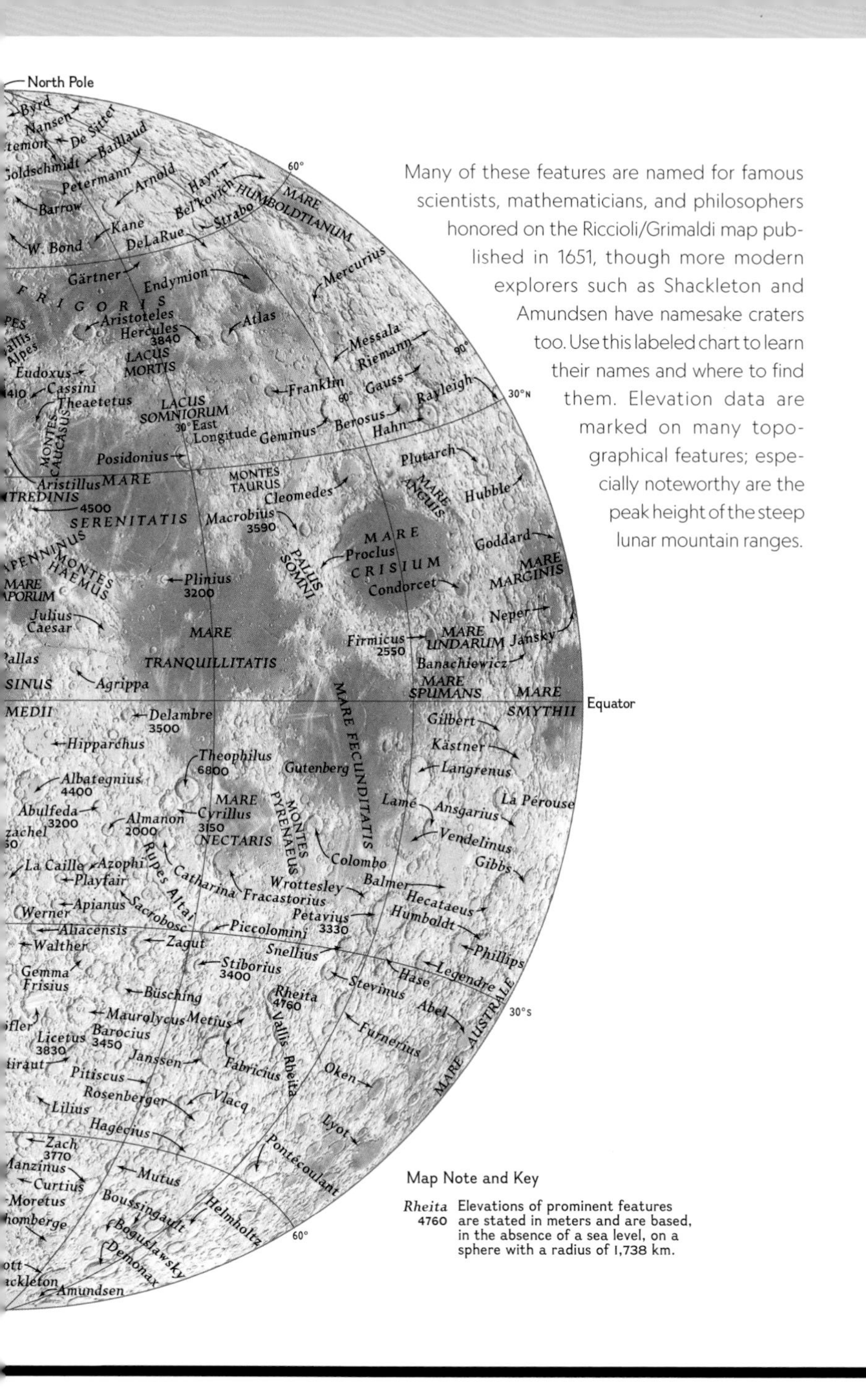

Many of these features are named for famous scientists, mathematicians, and philosophers honored on the Riccioli/Grimaldi map published in 1651, though more modern explorers such as Shackleton and Amundsen have namesake craters too. Use this labeled chart to learn their names and where to find them. Elevation data are marked on many topographical features; especially noteworthy are the peak height of the steep lunar mountain ranges.

Map Note and Key

Rheita 4760 Elevations of prominent features are stated in meters and are based, in the absence of a sea level, on a sphere with a radius of 1,738 km.

MOONGAZING

Lunar craters and mountains are visible through a telescope.

SEASONED AMATEUR ASTRONOMERS may find the bright moon a nuisance, but it offers plenty of amazing sights to explore. In fact, the moon is the best place to start observing with binoculars and small telescopes, especially where light pollution is an issue. You don't have to be Neil Armstrong to take an up-close-and-personal tour of the moon, either.

Binoculars

While not as powerful as a telescope, even a standard 7 x 50 pair of binoculars will reveal a plethora of the moon's intriguing features. Start by observing its changing lunar phases over time: The shifting illuminated portion of its disk can reveal a surprising number of details as lunar day turns to lunar night. Scan along the lunar terminator, the crisp boundary line between daylight and darkness, to reveal the sharply angled, long shadows cast by high mountain ranges and jagged peaks around the rim and center of craters. When the moon is near or at full phase, look for the bright dots of small craters scattered across the dark, gray maria. Steady views can reveal craters as small as tens of miles across. Look for larger craters like Tycho and Copernicus that sport bright, feathery, ray-like features radiating out in all directions.

FURTHER

Inconsistencies in the moon's orbit and rotation mean that nearby 60 percent of its surface is visible at one time or another.

As an observing bonus on certain nights, you can even catch the moon posing in your binoculars with bright planets, stars, and clusters. Occasionally, there are lunar occultations where the moon passes in front of deep-sky objects, with an effect similar to an eclipse. During a "grazing" occultation, the light of the distant star or planet will appear to wink on and off as the star light grazes off the visible edge of the moon. Exact locations and timings of the occultations, are provided by the International Occultation Timing Association.

Telescopes

A quick glance through even the smallest telescope reveals the difference in crater density between the much older, brighter highlands and the relatively young, dark maria. The best time to catch sight of most lunar features is not during full moon, when they appear whitewashed, but when cycling through its other phases. When it's a crescent, scan for the ever-changing lunar terminator. Magnification limits can range from 80x to 250x depending on atmospheric viewing conditions. From terraced craters and valleys to mountain ranges and escarpments, there are enough features here to turn the most grounded sky-watcher into an avid lunatic.

Venus passes behind the moon during a lunar occultation.

FURTHER

When the moon reaches first quarter phase every month, a weird topographical formation—a tiny letter X—magically appears through telescopes, and even binoculars, for about four hours. The walls of three closely clustered craters create this optical illusion when the crater floors are in total darkness and sunlight hits the craters' rims at just the right angle. Start your hunt for Lunar X about a third of the way up from the heavily cratered, southern face of the moon, along the terminator.

ARTIFICIAL SATELLITES

STEP OUTSIDE on a clear evening and you have an excellent chance of seeing a satellite glide through the sky. Keen-eyed observers can expect to see one every 15 minutes or so. The best satellite-spotting technique is to lie on the ground or in a reclining lawn chair and look up—no telescope or binoculars needed. You will know you're looking at a satellite, and not an airplane, by how it flies in a straight line and doesn't have flashing or colored lights.

Satellite Trackers

Those eager to find out exactly when to catch sight of a satellite will find convenient prediction services on the web. Generally geared toward sky-watching beginners, they allow you to track dozens of the brighter satellites and ascertain their predicted visible pass for your particular location. A couple of the most popular include *heavens-above.com* or NASA's *spaceweather.com*. By providing your city name or postal code, they offer up a list of suggested spotting times along with compass directions and altitude. Predictions computed a few days ahead are usually accurate within a few minutes. But they can change, owing to the slow decay of a satellite's orbit and periodic boosts to higher altitudes, as sometimes happens with space stations. Make sure to check again on the night of the flyby for the latest and most accurate predictions.

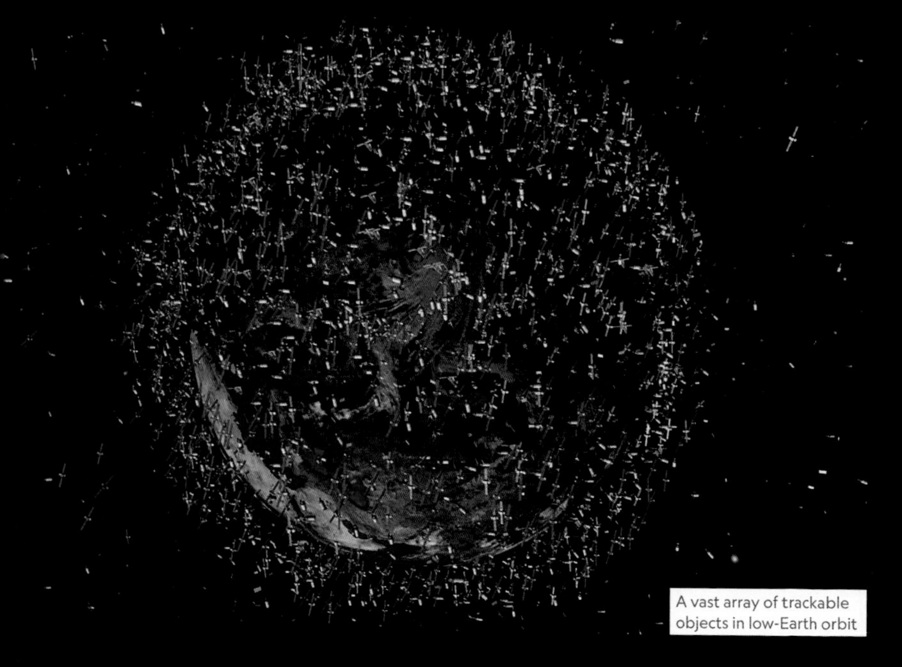

A vast array of trackable objects in low-Earth orbit

Earth Observation Satellites

In general, the best time to hunt for satellites is an hour or so after sunset. While Earth begins to cast its dark shadow around you, sunlight can still reach the high altitude of an orbiting satellite to bounce off its mirror-like solar panels or shiny metallic surface and make it visible from the ground as a traveling "star." The same thing happens when we see a high-flying airplane brightly lit by the recently set sun. After a few minutes, the satellite will enter Earth's shadow and appear to instantly vanish from view.

Surprisingly, you can spot quite a few Earth-observing spy satellites—but of course their identities are hard to pin down. Generally found at low altitudes, they zip across the sky in a north-south direction, visible only for a minute or two. Their high-speed polar orbits allow complete coverage of Earth over multiple passes.

Probably the worst-kept spy satellite "secret" is the highly visible Lacrosse series U.S. military radar-imaging satellites. Traveling quickly, and usually high overhead, these probes look like orange-hued stars because of their reflective orange insulation.

Transit of the swift International Space Station

Space Junk

Some of the artificial lights gliding above you may appear to twinkle as a retired satellite tumbles in its orbit, which means you've most probably locked onto some orbital debris, or "space junk." Space junk can be anything from defunct spacecraft to small pieces of broken human-made objects, traveling at speeds up to 17,500 miles per hour (28,200 km/h), which can pose a real threat to crewed space missions. Like satellites, space junk will disappear from view as it enters Earth's shadow on approach to the horizon, over a two- to five-minute period. That's very different from a falling meteor, which zips across the sky in a matter of seconds. NORAD Space Command tracks as many as 500,000 objects polluting the space around our planet, ranging from bus-size rocket boosters to small screws or hammers that floated away from astronauts during space walks.

FURTHER

Satellites orbit on an invariable schedule, so it is simple to predict their next visible flyby. Simply type your location into a satellite-tracking website to find the time it will pass overhead. Here are some recommended sites:

- *satview.org*
- *heavens-above.com*
- *spaceweather.com*
- *spotthestation.nasa .gov*
- *isstracker.com*

SPACE STATIONS

THE WORD "SATELLITE" can mean any object that orbits a planet. Our moon, for example, is Earth's natural satellite. The first artificial satellite was Sputnik I, launched into orbit on October 4, 1957; the first American satellite, Explorer I, would follow on January 31, 1958. Sputnik I marks the start of the space race, an intense competition between the U.S.S.R. and the United States. In 1971, the U.S.S.R. was the first to operate a space station, a large satellite where people can live and work for extended periods. Today the two nations are united in space: The International Space Station (ISS) is a joint effort between the United States, Russia, Japan, Canada, and 11 European countries.

Space stations are the easiest satellites to spot, since they are large, are highly reflective, and orbit at altitudes low enough that they appear bright to the naked eye. On a clear night, you can spot them even amid brightly lit city skies.

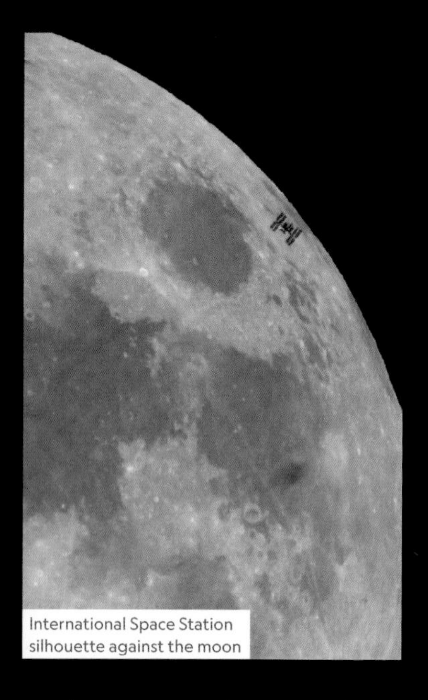

International Space Station silhouette against the moon

The International Space Station

Cruising at an altitude of about 280 miles (450 km), traveling at 17,000 miles per hour (27,360 km/h), and with a crew of six astronauts, this football field–size space station looks like a brilliant star to the naked eye. It moves at a good clip across the entire overhead sky, from eastern horizon to western horizon in two to four minutes, so you have to be quick when you are searching the skies.

In summer, usually around mid-June, observers in the northern latitudes get a great chance to catch sight of the ISS as it makes not just one, but multiple flybys. Its 90-minute-long loop around the planet closely traces Earth's day-night border, so it is bathed in perpetual sunlight. The ISS is lit from dusk until dawn; observers are treated to a veritable station-spotting marathon, and anyone with clear skies can observe it up to five times in a single night. That's impressive when compared with other times of year, when it's sunlit only about 70 percent of the time and observers can spot it only once or twice a night.

Chinese Space Stations

Chinese space station Tiangong-2, or Heavenly Palace 2, launched in 2016, was designed to test capabilities for long-term human presence in anticipation of the nation's permanent space station, which should go into orbit by 2024. The station can vary in brightness from a very faint magnitude 4 to a bright, naked eye magnitude 1, making it a fairly easy sky-watching target. China's earlier generation station Tiangong-1 was decommissioned and sent hurtling toward Earth in April 2018, burning up as it entered the atmosphere.

Photographing Transits

Luckily for the avid sky-watcher, advances in technology mean that you can view, track, and photograph the ISS through an amateur telescope or a camera with a long enough focal length. Eight- to 10-inch telescopes will let you view some of the station's structural details, such as its solar panels and docked cargo ships. Some clever backyard astronomers have even figured out how to capture snapshots of the satellite as it glides directly in front of the full moon or sun.

To do this, you need the right equipment: a computer-controlled telescope or camera and some satellite-tracking software like that at *www.transit-finder.com*. When you choose a specific satellite from the database, the telescope will move to the correct position, lock itself onto the satellite, and follow it until it disappears over the horizon. Timing is everything, since the ISS takes as little as a third of a second to move across the sun's or moon's disk. The path of visibility is extremely narrow at only a few miles across, making capturing the ISS even trickier. Your best hope is to take a quick sequence of pictures as it zips across the lunar or solar disk.

A spacecraft docks at the International Space Station.

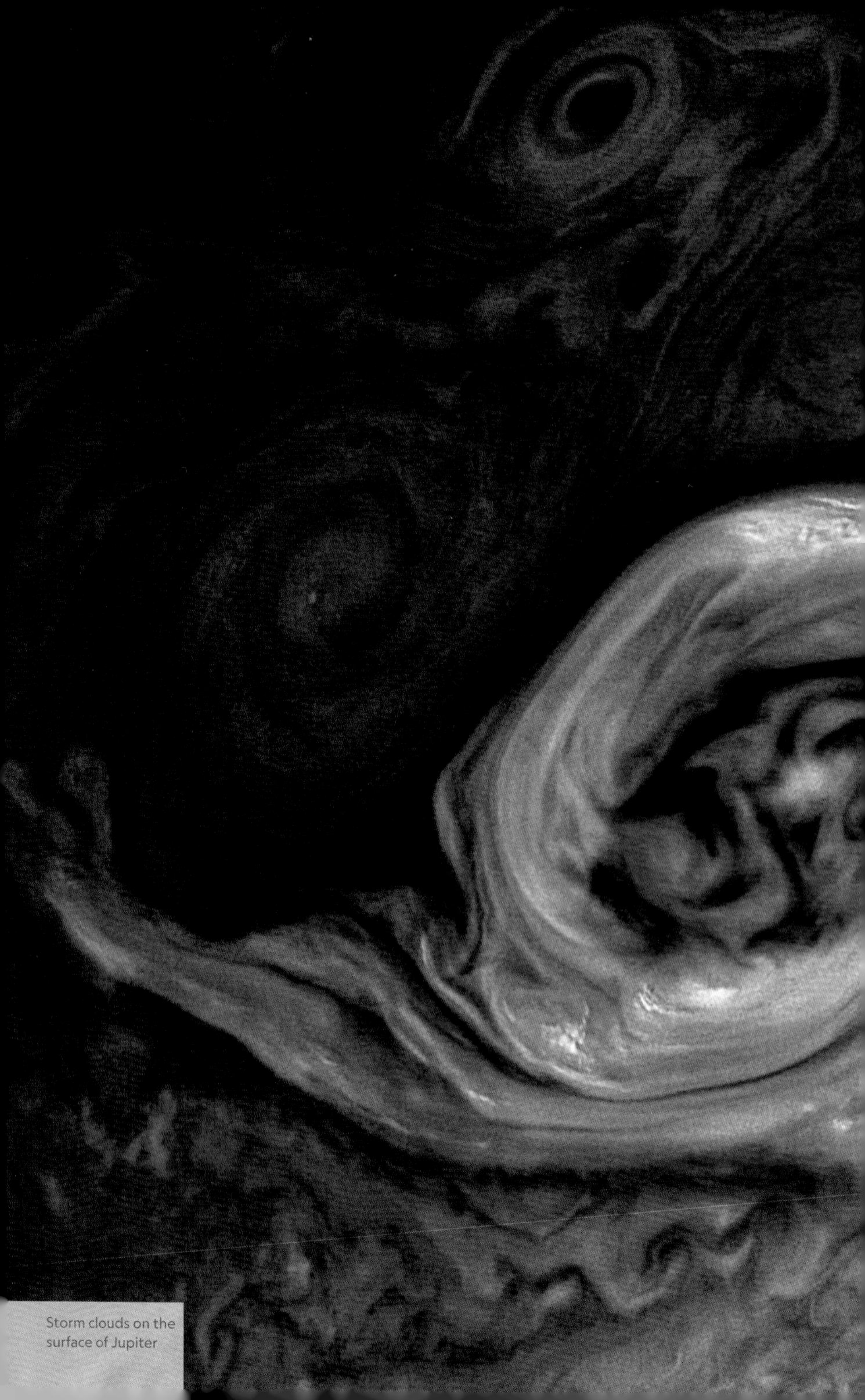

Storm clouds on the surface of Jupiter

CHAPTER 5

THE PLANETS

THE SOLAR SYSTEM

FURTHER

Need a way to remember the order of the planets in the solar system (Mercury, Venus, Earth, Mars, Jupiter, Saturn, Uranus, Neptune)? The International Astronomical Union (IAU) suggests using this mnemonic phrase: "My very educated mother just served us nachos."

THE SOLAR SYSTEM is dominated—as the name implies—by the sun. The sun's gravity holds the structure in place, fueling life on Earth with its energy, and the sun accounts for all but a minuscule portion (about 0.2 percent) of the solar system's mass. But that tiny fraction still includes a diverse array of celestial objects, spread out over a distance of about 100,000 Astronomical Units (AUs)—from the sizzling innermost planets, through a disk of dust scattered across interplanetary space, to the frozen outer reaches.

Planets

The most easily located objects in our solar system (excluding Earth, the sun, and the moon, of course) are the planets. They are spread across a distance ranging from Mercury, at roughly 36 million miles (58 million km) from the sun, to Neptune, whose orbit keeps it at an average distance of around 2.8 billion miles (4.5 billion km) from our central star. Because of the dynamics of the solar system's formation, all major planets travel close to the ecliptic, along the plane that was laid out by the spinning motion of the gas cloud that formed the sun. Following along with them—particularly with gas giants like Jupiter—are dozens of moons, many of which are good telescope targets and some of which can be spotted with binoculars. The two innermost planets are unable to stably hold their own moons owing to their proximity to the sun's gravity.

Worldlets (Comets, Asteroids & Meteors)

At the outer edge of the solar system is the distant Oort cloud, home to billions of massive, drifting ice balls that, when dislodged, can speed through the solar system as fiery comets. The Oort cloud is one of three of our solar system's comet breeding grounds. The other ones lie much closer, in the Kuiper belt and its outer extension, the scattered disk. The Kuiper belt and scattered disk are regions of icy, rocky debris and near-planet-size objects that begin just beyond Neptune's orbit and extend to about 1,000 AUs.

Closer in, a 340-million-mile (547 million km) gap between Mars and Jupiter is littered with asteroids, rocky fragments left over from the solar system's early days. This "asteroid belt" marks the border between the terrestrial planets and the gas giants. Its existence is attributed to Jupiter's massive gravitational field, which prevented any other planet from coalescing in that space and left scattered bits of debris to collide into and pulverize each other, creating ever smaller pieces.

The brightest asteroids can be spotted using binoculars. More often, they are noticed when smaller ones come burning through Earth's atmosphere as meteors, a catchall term for falling interstellar debris and more commonly known as shooting stars. Those that land on Earth are known as meteorites.

SKY-WATCHERS
Johannes Kepler

German mathematician and astronomer Johannes Kepler offered a bridge between the lingering classical notions of a solar system, built from perfectly ordered spheres and circular motions, and the more chaotic reality that appeared as the telescope and subsequent inventions allowed for better observation. In the late 16th and early 17th centuries, Kepler set out to prove that the underlying motions of the planets did follow a pattern—even though it had become apparent that they were not moving in perfect circles as assumed by earlier astronomers. He ultimately formulated three laws that describe the elliptical motion of celestial bodies around each other—rules that explained what was happening with the planets but that also apply to moons, comets, and other objects. Even as the cosmos became more crowded, the underlying order of Kepler's laws of planetary motion continued to apply.

WHAT IS A PLANET?

THE ANCIENT GREEKS wondered why the asters planetes, or "wandering stars," seemed to saunter around the sky, sometimes disappearing from the sky altogether, then returning at another time of year. The Romans thought these brilliant points of light were deities, and their names are still with us today—Mars the god of war and Venus the goddess of love. Had the Greeks known about Jupiter's dozens of moons or the icy detritus circling in the distant Kuiper belt, they may have given up their notions of the universe being a polished and perfectly ordered sphere.

Planetary Qualifications

We now know that planets are not stars and their twinkle is the reflection of sunlight. The visible changes in a planet's apparent position, relative to the stars, result from its orbit around the sun. Recent debate over the status of Pluto, which is no longer considered a planet, also produced a new definition for what separates these wandering bodies from other objects in the solar system.

FURTHER

When three celestial bodies array in a straight line, as during a solar or lunar eclipse, it is called opposition or conjunction. A more general term for it is syzygy.

In the early 21st century, discoveries of planetlike objects in the distant Kuiper belt caused the IAU to set three criteria for official planetary status. First, planets are bodies that orbit the sun. Second, they have a roughly round shape as a result of their own gravity. And third, they are large enough to have cleared their orbital path of debris. As a result, Pluto was demoted to a dwarf planet: Though planetlike, it hasn't cleared its orbit of stray solar system debris.

Yet those criteria have not settled all the mystery surrounding Earth, its planetary companions, their satellites and rings, and the other residents of the solar system. Though as many as 10 percent of stars have at least one planet

Planet Earth

Illustration of an exoplanet orbiting its star

in orbit, neighborhoods as crowded as ours may prove to be rare.

Beginnings

Most of the objects in our solar system share an origin: They came from the nebulous solar cloud that eventually coalesced into the sun. As the cloud rotated, it developed a round embryonic sun surrounded by a wide, flattened disk of dusty gases. Pressure increased in this proto-sun's heart, triggering the nuclear fusion that set the sun's life cycle in motion.

The material left circling around the sun, meanwhile, began to evolve into separate bodies. Bits of rock and ice began sticking together and accumulating as planetesimals, gathering mass and eventually building up to planetary size. Clouds of gas were drawn into orbit around them. Closer to the sun, the objects tended to be denser and rockier, with lighter atmospheres or none present at all. These formed what are now known as the terrestrial planets: Mercury, Venus, Earth, and Mars. Farther away, where the sun's vaporizing effects were less profound, large clouds of gas collected around cores of rock and ice, forming the planets of the outer solar system. Despite their large size, gas giants Jupiter and Saturn are composed of the lightest of gasses enclosing small rocky or liquid cores. At the far edge of the solar system, the ice giants Uranus and Neptune are composed of slightly heavier gases.

THE SCIENCE OF Exoplanets

We have cataloged some 4,000 exoplanets—worlds circling stars other than the sun. The deeper we peer into space with technology like the Kepler Space Telescope, the more we find—enough to raise doubts about whether our solar system is as unique as we think. Astronomers estimate that the Milky Way may be home to tens of billions of Earthlike worlds with potential to hold life.

VIEWING THE PLANETS

FURTHER

Other planets—and even some of their moons—also have auroras, just like the ones at Earth's poles.

OBSERVING THE PLANETS involves a wide range of effort and equipment, depending on the target and the level of detail you're aiming for. Venus—the brightest object in the sky after the sun and moon—can be glimpsed with the naked eye, as can Jupiter. Saturn can be seen unaided, but spotting its rings requires a telescope with an aperture of more than 3 inches (76 mm); Jupiter's moons appear as starlike points through binoculars, while a telescope with an aperture of at least 2.5 inches (63.5 mm) will display the planetary disk flanked by its satellites. Finding distant Neptune will require a telescope, chart or software, and practice in discerning it from its neighbors.

Planet-Finding Resources

Charts, almanacs, and other resources are helpful for determining the best planetary viewing times. Published resources will typically place the planets month to month in different constellations, mostly along the zodiac. Turn to the appendix on page 276 to locate planets in this manner. Special events will also be featured in places like *Sky & Telescope* magazine and *Astronomy* magazine's online "StarDome."

Rules of Thumb

You can spot some planets without the help of gear or charts—if you know where and when to look for them.

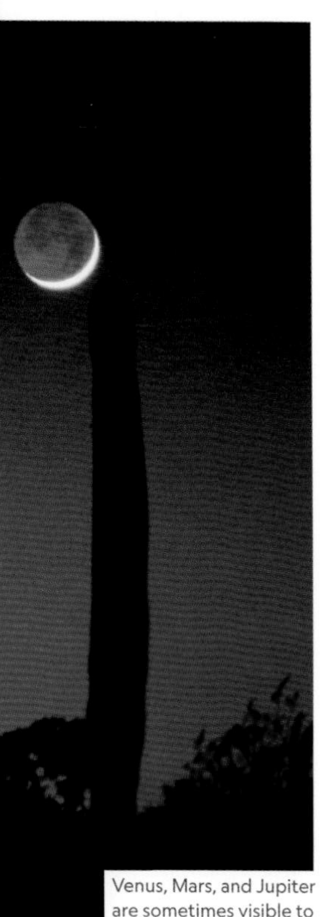

Venus, Mars, and Jupiter are sometimes visible to the naked eye.

Mercury, the planet closest to the sun, is visible only at dusk and dawn. Much brighter than Mercury, Venus gets higher in the eastern and western skies.

Mercury and Venus are interior planets, so they disappear twice in their orbital year—once when they pass between Earth and the sun (a point called inferior conjunction) and once when they pass a point that is, from Earth's perspective, behind the sun (called superior conjunction). The exterior planets, from Mars and beyond, can appear anywhere along the ecliptic. The best viewing is when Earth passes between the planet and the sun, a situation called opposition, when they are bathed in direct sunlight and will be visible almost all night. Opposition is also what causes the full moon. It happens roughly once a year except for Mars, as its orbital period is much closer to Earth's and thus much harder for us to catch.

THE STORY OF Planet Vulcan

Neptune's discovery in 1846 began the hunt for another planet: one that was thought to be causing changes in Mercury's orbit. Urbain Jean Joseph Le Verrier, one of the astronomers who had foreseen Neptune's presence, calculated that a planet about the size of Mercury and orbiting half as close to the sun might be the reason for Mercury's deviation. In 1859, he announced that he had discovered a new world, called Vulcan, speeding around the sun every 20 days. What had been observed turned out to be a sunspot. It took Einstein's theory of general relativity, developed early in the 20th century, to explain the slippage in Mercury's orbit.

MERCURY

SYMBOL: ☿

RADIUS: 1,516.0 mi (2,439.7 km)

MASS: 0.06 x Earth's

DISTANCE FROM SUN: 35,983,095 mi (57,909,175 km)

LENGTH OF YEAR: 88 Earth days

LENGTH OF DAY: 58.6 Earth days

SATELLITES: none

MAGNITUDE IN SKY: up to -1.9

THIS SMALLEST AND INNERMOST PLANET is close enough to be seen with the naked eye, but is also one of the most difficult to spot. It lies less than half the distance between Earth and the sun—about 0.39 AU, or 36 million miles (58 million km)—and circles the star every 88 days, which means it's frequently hidden from view in the glare of sunlight. It is visible six times in a typical year, either low in the west after sunset or low in the east before sunrise. Mercury is small, and getting smaller: Studies of its crust suggest it has shrunk by as much as 9 miles (14 km) over its history. A barrage of solar wind and micrometeorites kicks up surface material that Mercury's weak gravity can't keep from floating off into space, creating a comet-like tail behind the planet.

On the Planet

Little was known about the surface of Mercury until the mid-1970s, when photographs from the Mariner 10 spacecraft showed us a barren, rocky planet strewn with craters, layered in a way unlike anywhere else. It has, in essence, no atmosphere and consequently is a land of temperature extremes. The region around the equator may range from

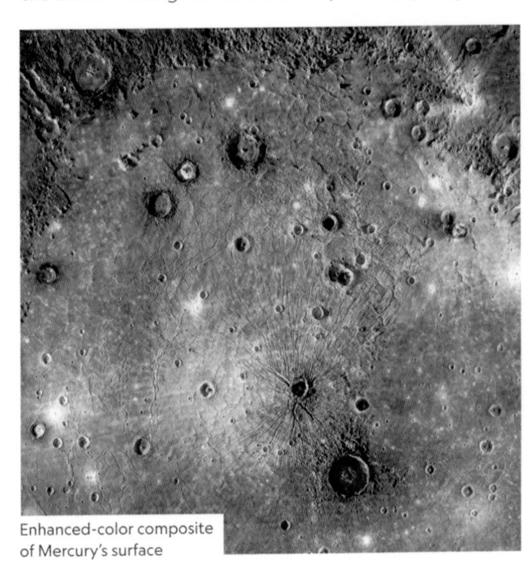

Enhanced-color composite of Mercury's surface

FURTHER

While orbiting Mercury from 2011 to 2015, NASA's MESSENGER probe revealed Mercury's surface composition and geological history; discovered its internal magnetic field is offset from the planet's center; and proved that its polar deposits are predominantly water ice.

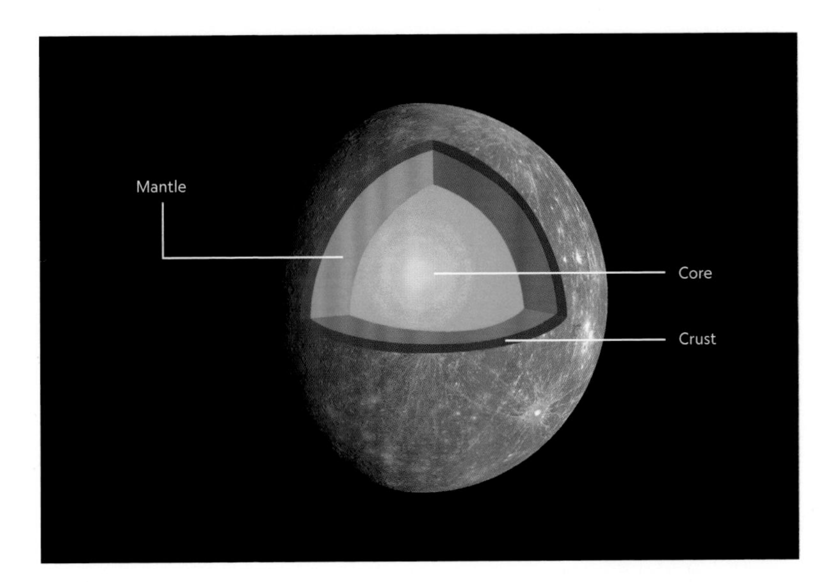

800°F (427°C) during the day to minus 300°F (−184°C) at night. Its massive iron core makes up about 65 percent of its mass.

Viewing Mercury

The best times to hunt for Mercury are during the fall and spring, when the ecliptic stands at its sharpest angle relative to the horizon and the tiny planet reaches its widest separation from the sun. For Northern Hemisphere observers, evening in March and April and morning in September and October are the prime viewing times.

If the atmosphere cooperates and the horizon is clear and unobstructed, the planet might be seen on consecutive nights for as long as three weeks running before it disappears into one of its frequent conjunctions with the sun. But turbulence is more likely to occur near the horizon, so the image may prove little better than a shaky speck. Similar to Venus, it will appear to go through phases—gibbous, half-lit, then crescent—as it moves around the sun and catches up with Earth. Binoculars are useful for locating Mercury in a twilit sky, but they won't show its phases as a telescope can.

FURTHER

In Roman mythology, Mercury is the fleet-footed messenger of the gods, and his planetary namesake clips along at nearly 31 miles (50 km) per second faster than any other planet. Your best bet for seeing Mercury in the dusk or dawn skies is to find a viewing location with an unobstructed view of the horizon and sweep the low sky with binoculars.

VENUS

SYMBOL: ♀

RADIUS: 3,760.4 mi (6,051.8 km)

MASS: 0.82 X Earth's

DISTANCE FROM SUN: 67,238,250 mi (108,209,475 km)

LENGTH OF YEAR: 225 Earth days

LENGTH OF DAY: 243 Earth days

SATELLITES: none

MAGNITUDE IN SKY: up to −4.5

FURTHER

At its brightest, Venus is eight times brighter than Jupiter and 23 times brighter than Mars, making it an easy target for first-time viewing.

Venus

WITH A DIAMETER AND MASS similar to Earth's and a 225-Earth-day orbit around the sun, Venus was long thought of as something of a sister planet—perhaps one with a lush, tropical climate just waiting to be discovered. But this planet named for the Roman goddess of love has hardly proved welcoming.

On the Planet

Venus is the hottest planet in the solar system, with a surface temperature in excess of 860°F (460°C). Though farther from the sun than Mercury, Venus's surface temperature doesn't change from day to night or pole to pole. That's because of a runaway greenhouse effect: Thick with carbon dioxide, the planet's dense atmosphere traps the sun's heat.

Venus is blanketed in clouds of sulfuric acid, a highly reflective cover that makes it shine brightly while also hiding its surface from view. Our knowledge about its surface conditions has all come from space probes, beginning with the U.S. Mariner 2 in 1962. Its atmosphere whips around it in just 96 hours, creating massive, shape-shifting vortices at its poles. Metallic "snow" falls onto the volcanoes pockmarking its surface, where the air weighs roughly 90 times more than on Earth.

Viewing Venus

As one of the two interior planets, Venus will always be

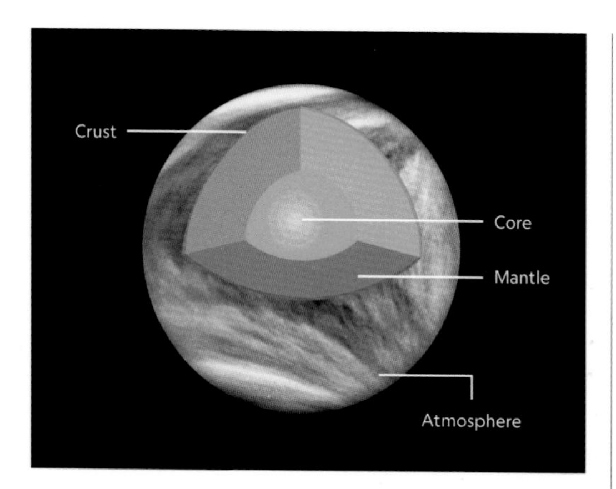

seen in the hours after sunset and before sunrise; it never moves too far from the sun. Because of its proximity to Earth, and its similar orbital year, it is frequently visible for at least some part of the night.

It will hang low to the horizon at nightfall in the western sky, appearing at its smallest and dimmest. As it continues on its path, Venus moves farther from the sun—from an earthbound perspective—and higher in the sky. When it arrives at its greatest eastern elongation (the greatest angular separation from the sun), it will be at its highest point, have more than doubled in apparent size, and reached it brightest magnitude of -4.5.

Venus grows larger and brighter as it gradually overtakes Earth from behind. To the unaided eye, that brightness will mask the fact that Venus moves through phases similar to that of the moon. At its two elongations, it is half lit. While approaching inferior conjunction, roughly every 18 months, it enters a crescent phase that can be seen through binoculars (and, for those with excellent vision, with the naked eye), before disappearing as it passes between Earth and the sun.

When Venus overtakes Earth in its orbit, it becomes the morning star, rising just before the sun, at first, but appearing a little earlier and higher each evening. Eventually, it disappears into superior conjunction before a new apparition begins.

THE STORY OF the Transit of Venus

About four times every 243 years, Venus's orbit takes it across the face of the sun. The most recent transit was in 2012, and it won't happen again until 2117. In the 1760s, one of these rare anticipated transits inspired England's Royal Society to send a team to observe it from the southern latitudes, and naval authorities appointed James Cook to captain the H.M.S. *Endeavour*. He was also tasked with finding a southern landmass that was assumed to exist there, which led him to chart New Zealand and parts of Australia.

Venus transiting the sun

MARS

SYMBOL: ♂

RADIUS: 2,106 mi (3,389.3 km)

MASS: 0.11 x Earth's

DISTANCE FROM SUN: 141,637,728 mi (227,943,827 km)

LENGTH OF YEAR: 687 Earth years

LENGTH OF DAY: 24.6 Earth days

SATELLITES: 2

MAGNITUDE IN SKY: +1.8 to −2.91

A MANNED TRIP TO MARS remains a goal for many of the world's national space agencies. The danger and logistics of such a mission—with round-trips of as much as 900 days—make it a voyage that is still most likely a couple of decades off. But with more than 50 unmanned missions launched or attempted since the 1960s, Mars is already one of the most frequently visited planets.

On the Planet

The first flyby probe, Mariner 4, which reached Mars in 1965, revealed a barren, rocky landscape with a dusty surface made mostly of silicon, sulfur, and the iron oxide that accounts for its rusty color. The atmosphere is thin—less than one percent of Earth's and made mostly of carbon dioxide—and cold, with average temperatures of minus 85°F (−65°C). The south is dominated by cratered highlands and the north by smoother plains. Like Earth, Mars has seasons and weather. Dust storms and clouds move across the planet, and its frozen poles are capped with ice that shrinks and grows as the seasons change.

With a planetwide desert and, in some areas, windswept dunes, the red planet claims some of the solar system's most dramatic topography. That includes an

Mars

FURTHER

Mars's two moons, Phobos and Deimos, bear names from the Greek for "fear" and "flight" that echo Mars's forceful demeanor.

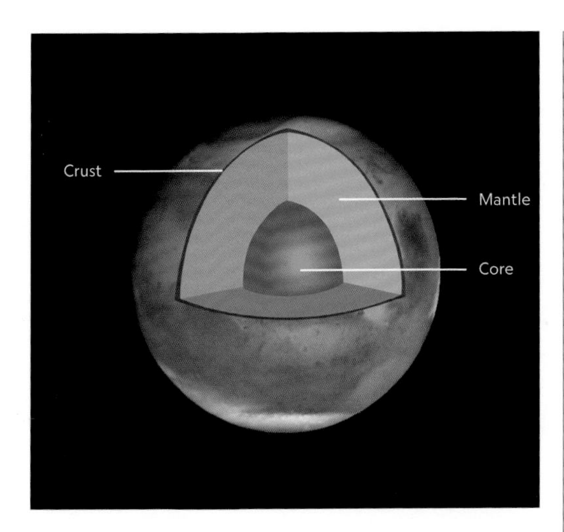

equatorial region called Tharsis, which is home to some truly massive volcanoes. With no plate tectonic movement, many have grown to enormous sizes as lava layers flow over each other in the same hot spot over billions of years. Olympus Mons, the largest volcano in the solar system, rises some 16 miles (26 km): three times higher than Mount Everest. Scientists believe that while most have gone extinct, a few show tantalizing signs of recent activity.

Also located near the equator is the impressive Valles Marineris, a vast canyon network that spans about 2,500 miles (4,000 km). Arizona's Grand Canyon would easily fit inside one of its minor tributary fissures.

With mounting evidence for extensive flood plains and shallow seas, theories now suggest that four billion years ago Mars may have been much more Earthlike, with a much thicker atmosphere and abundant surface water. Rovers have found features such as sedimentary rocks and minerals that could only have been formed by water over long periods of time. But if so, what happened to it? Without any active geological processes, the planet's atmosphere may have gotten colder and thinner. Water may have seeped into the ground and formed frozen reservoirs, which could account for why present-day Mars is mostly a frozen, dry world.

SKY-WATCHERS
Percival Lowell

American astronomer Percival Lowell is known for one of modern astronomy's most imaginative miscalculations. He was a chief proponent of the idea that faint linear markings on Mars's surface were evidence of an irrigation system built by an advanced civilization. Though the idea was never widely accepted (and finally laid to rest when Mariner 4 sent back the first photos of the barren planet in 1965), Lowell publicized his theories in a book, *Mars and Its Canals*. Why canals? The theory may have started with an overly aggressive translation of works by Italian astronomer Giovanni Schiaparelli, who in 1877 spotted lattice-like lines he referred to as *canali*, by which he meant natural channels in the landscape, not human-made structures.

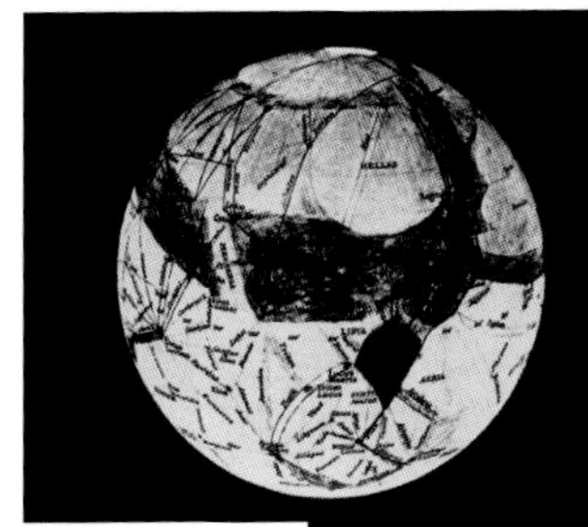

Lowell's map of Mars

MARS UP CLOSE

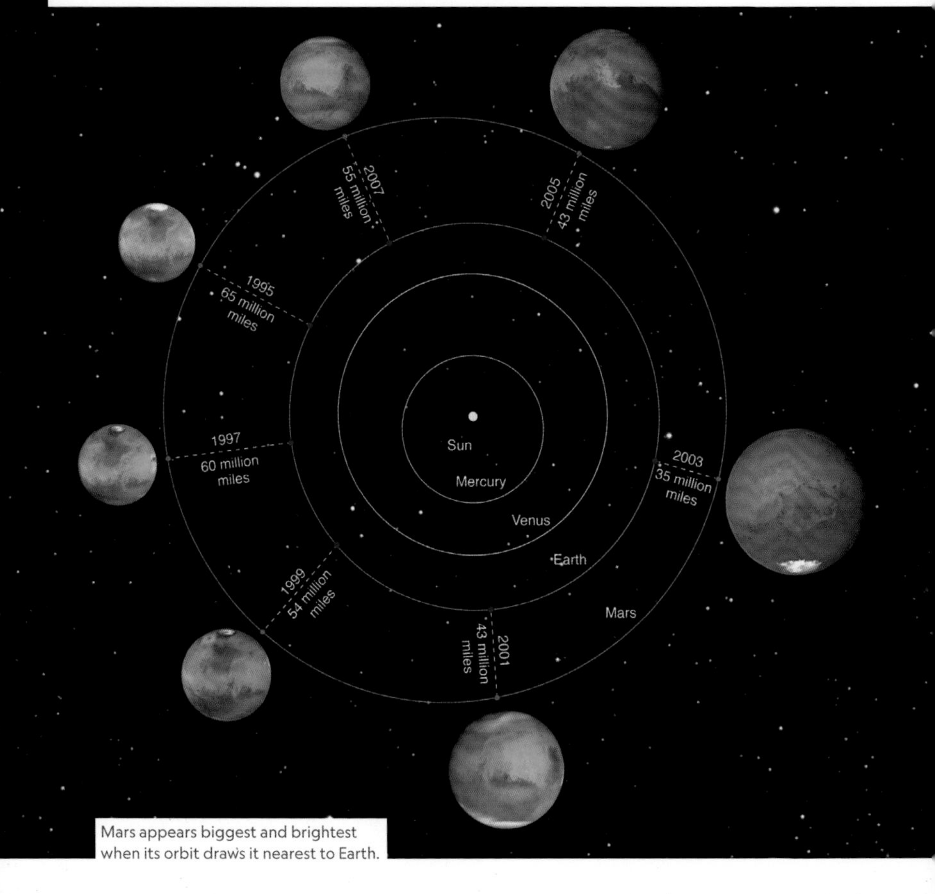

Mars appears biggest and brightest when its orbit draws it nearest to Earth.

FURTHER

Enhance views of Mars's darker markings by using orange or red eyepiece filters. Pale-blue-colored filters will bring out its white polar caps and clouds.

IN MANY WAYS, Mars is Earth's closest cousin. It is tilted similarly to the ecliptic plane and thus experiences seasons and weather. Although Mars's rotation axis tips about as much as Earth's does, its orbit is elliptical and angled rather than circular. As a result, its seasons aren't the same length as each other or the same in each hemisphere.

Viewing Mars

Mars can be seen with the naked eye, but its apparent size and brightness vary markedly during its 687-day orbit. The naked eye will show its ruddy color, but through a telescope of modest size—a 4-inch (100 mm) refractor or 6-inch (150 mm) reflector—you will be able to glimpse distinct dark and light markings, depending on which side of the planet is facing Earth. Explore dark features like Syrtis Major or the Hellas Basin, as well as the white, frozen polar caps and even some water-ice clouds. At times, there are even opportunities to watch a moving dust storm.

Timing Your Observation

You'll get the best views of Mars at opposition—when Earth sits between it and the sun and the disk appears at its biggest and brightest. Because its orbit closely matches Earth's, those moments are infrequent compared with the roughly annual oppositions of the gas giants. From one opposition of Mars to the next takes roughly 780 days, and not all oppositions are created equal. The red planet's eccentric, stretched-out path brings it as close as 35 million miles (56 million km) to Earth roughly every 17 years, when Mars is both at perihelion (closest point to the sun).

At one point in Mars's journey across the sky, it will appear to briefly stop, then move backward in a phenomenon called retrograde motion. Mars begins each new apparition near the eastern horizon, rising before sunrise and moving progressively farther from the sun and higher in the sky. As Earth overtakes it in orbit, Mars's motion appears to stop before it begins a roughly three-month journey westward. After coming into opposition, it will resume an eastward course, appearing ever lower in the evening sky before slipping into the sun's glare.

THE SCIENCE OF Martian Life

To date, no hard evidence of past or present life on Mars has been found. However, after landing in 2012 in what was once thought to be a giant, lake-filled crater fed by streams billions of years ago, NASA's Curiosity rover found minerals, clay, and conglomerate rocks like ones found on Earth wherever water flows over extended periods. Researchers today believe Gale crater was a lake, most likely awash with the basic chemical ingredients needed to create a veritable utopia for simple life-forms.

Self-portrait of the Mars Curiosity rover

ASTEROIDS & DWARFS

BETWEEN MARS and Jupiter, the territory that divides the terrestrial planets and the gas giants is populated by millions of rocky objects. These minor planets or asteroids (from the Greek *aster*, or "star," and *-oid*, meaning "like") never coalesced into a planet because of the influence of massive Jupiter.

Asteroids Abound

Of the profusion of asteroids in space, more than 780,000 have been cataloged, the vast majority in the asteroid belt. More than 21,000 have been named—after an honor roll of girlfriends, religions, and favorite writers, among others. Each member of the Beatles has one named after him.

There may be millions of asteroids out there. Though they are too small to be spotted, their presence can be felt in other ways. Objects in the asteroid belt frequently collide, and their debris can come burning through Earth's atmosphere as a meteor. Occasionally they even reach Earth's surface as a meteorite.

In one of the small fields where asteroids have collected along Jupiter's orbital path, among the so-called Trojan asteroids, there is one large rock in the asteroid belt that stands out. With a diameter of 580 miles (933 km), Ceres is no longer considered an asteroid. It was given dwarf planet status after the IAU created that category in 2006 to define planetlike objects that share an orbit with others.

FURTHER

Some of the largest asteroids (Ceres, Vesta, Pallas, Juno), when at opposition, are bright enough to see with binoculars and even the naked eye. While not the biggest, 330-mile-wide (530 km) Vesta has such a highly reflective surface that it can shine brighter than the planet Uranus.

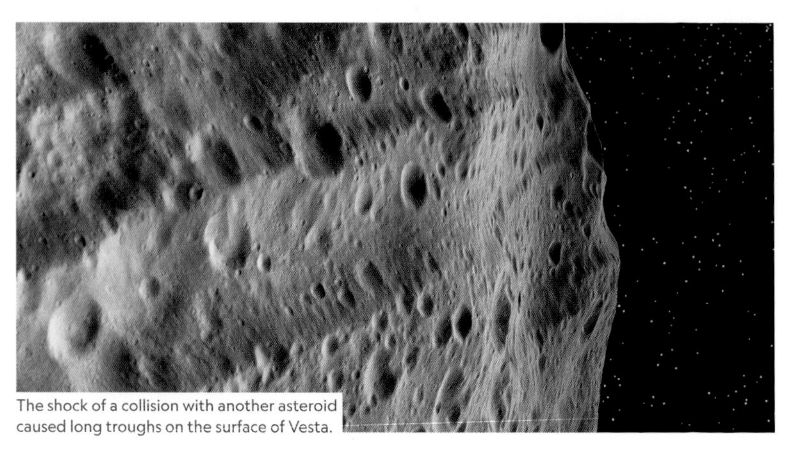

The shock of a collision with another asteroid caused long troughs on the surface of Vesta.

Dwarf planet Ceres

Seeing Asteroids

A few asteroids are visible with binoculars, while hundreds are theoretically within the range of even a modest 3-inch (76 mm) telescope. Finding them is another matter entirely. Astronomical publications will provide lists of dates and directions to the most prominent ones, known as ephemerides. Remember, these objects will be faint and largely indistinguishable from stars. There are two strategies for picking them out once you locate the right area in the sky. Using a low-powered eyepiece, you can map (or photograph) what you see and compare it with observations later in the evening or on a subsequent night to determine what's moved against the backdrop of "fixed" stars. Alternatively, you can compare the field of view with a sky chart: The asteroid will be the starlike object not on the map.

Observing Earth-crossing asteroids as they barnstorm Earth is a thrilling sky-watching experience. Despite being very faint, it's possible to detect an asteroid's motion with telescopes as they pass in front of fixed stars. The best bet for chasing them down is to keep an eye on late-breaking news of any impending close encounters from websites like *SkyandTelescope.com*, *spaceweather.com*, and *Astronomy.com*.

SKY-WATCHERS
Giuseppe Piazzi

Astronomers in the late 1700s were hunting for a planet in the space between Mars and Jupiter. It was an Italian monk who found what everyone was looking for—or so they thought. In 1801 Giuseppe Piazzi, observing at the Palermo Observatory in Sicily, located an object that was acting planetlike. It was dubbed Ceres, but doubts about its size, shape, and presence among many similar bodies eventually led people to classify it as the largest asteroid—until it was redesignated in 2006 as a dwarf planet.

JUPITER

SYMBOL: ♃

RADIUS: 43,441 mi (69,911 km)

MASS: 317.82 x Earth's

DISTANCE FROM SUN: 483,638,563 mi (778,340,821 km)

LENGTH OF YEAR: 12 Earth years

LENGTH OF DAY: 10 Earth hours

SATELLITES: 66 moons

MAGNITUDE IN SKY: −1.6 to −2.6

JUPITER IS ROUGHLY 15 TIMES FARTHER away from Earth than Venus is, but at its most brilliant it looms nearly as large and bright. Fortunately for sky-watchers, Jupiter's 12-Earth-year orbit allows it to spend around a year in each zodiac constellation. Jupiter governs its own miniature solar system of at least 79 moons. The four largest ones—Io, Europa, Ganymede, and Callisto—are visible with binoculars. First spotted by Galileo in 1610, they are referred to as the Galilean moons.

On the Planet

Jupiter is large enough to swallow 1,200 Earths—twice as massive as all the other planets combined. All told, its size, orbit, and constantly changing atmosphere make Jupiter one of the most popular planetary targets for sky-watchers.

With Jupiter, we never see a solid surface: just roiling clouds that are thousands of miles thick. Descending through Jupiter's atmosphere, intense pressure creates a zone of metallic hydrogen that surrounds the planet's rocky, molten iron core, the closest thing to a "surface." Currents swirling within this material create a large magnetic field, intense emissions of radio waves, and regular bursts of radiation. Jupiter traps electric particles streaming from both the sun and the moon Io, which create a magnetosphere with a shape similar to the Van Allen belts around Earth, but much more intense.

Jupiter

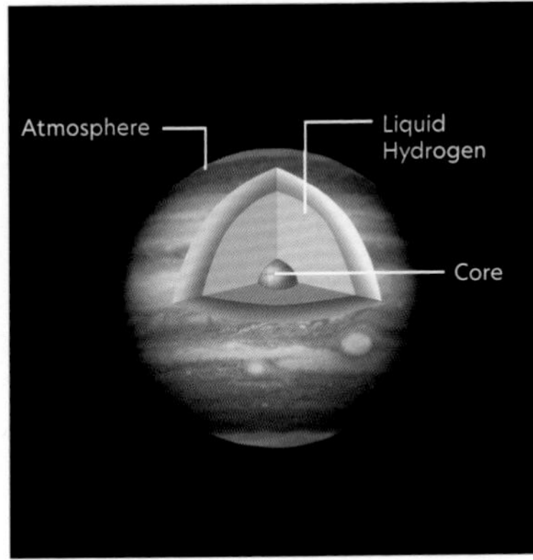

Jupiter shines brightly in a sky lit up by the Milky Way.

Viewing Jupiter

Jupiter appears as a superbright star to the naked eye, but a small telescope using only 20x magnification details will reveal its opaque atmosphere. Try a 6-inch (150 mm) scope or larger. Jupiter's quick 10-hour rotation creates swirling winds that make faint, pastel-colored bands in the planet's thick clouds of hydrogen, helium, methane, and ammonia. The most prominent features are two warmer, dark-brown stripes that appear on either side of the equator, known as Equatorial Belts. The lighter regions between the belts, known as zones, are clouds of crystallized ammonia that hide the darker lower layer from view. Where the belts and zones collide, massive cyclonic storms are born, visible at a higher magnification as dark or white spots.

These persistent storm systems only add to the spectacle. The most famous one is the Great Red Spot, a high-pressure zone that is about two times as big as Earth. The Great Red Spot—a circle-like patch below the equator—ebbs and flows but is typically visible. Even when the spot disappears its absence can be detected in the Red Spot Hollow, a bend in the adjoining cloud belt. With your telescope, a light-blue filter over the eyepiece will make the lines between the atmospheric bands more definite, while yellow and orange filters will help bring out other details.

FURTHER

Massive Jupiter falls short on one front. Like Saturn and Uranus, it has a ring, which was first revealed by the Voyager 1 probe. But that ring is small and thin compared with the ones around the other planets. It's only about 3,700 miles (6,000 km) wide—a puny crown for the planetary king.

JUPITER'S MOONS

Jupiter with its moons

KING OF ITS OWN miniature system of worlds, Jupiter truly lives up to its godly status by ruling over an entire family of moons. The four largest have become a focus for scientific exploration. Io is the most volcanically active world in the entire solar system, with hundreds of constant eruptions. Europa is encased with a fractured ice layer, which astronomers suspect hides a briny ocean. Ganymede is the largest moon in the solar system, and its ancient, crater-riddled surface may also conceal a subsurface ocean. Callisto is heavily cratered, like Ganymede.

Viewing the Galilean Moons

You can spot these larger moons with a pair of binoculars, which are bound to be far better than what Galileo was using when he first discovered them back in 1610. You can easily make out the outer moons with a 7x or 8x magnification, and using 10x or greater will bring all four Galilean moons into focus. Io being the closest to Jupiter will appear to move the most rapidly, making its movement readily visible within an hour or two.

Remember to hold your binoculars steady; better yet, get a binocular mount to prevent the shakes. Watch over the course of several nights and you'll notice that not all four moons are always visible. They appear in constant motion, changing positions as they eternally revolve around the planet in their orbits. As each moon circles around Jupiter, one or more may be directly in front or behind the planet. Backyard astronomers with medium-size telescopes between 6 and 8 inches (150 to 200 mm) can actually track the shadows of the moons on Jupiter's cloud tops.

Sometimes the orbital plane of Jupiter's Galilean moons will happen to appear edge-on from our line of sight, which means the moons can appear to eclipse their neighbors when their shadows are cast off their disks. Try sketching what you see, plotting the movements of the moons just as Galileo did. Draw the disk of the planet, then pinpoint the position of the moons on or around it, over a few subsequent nights.

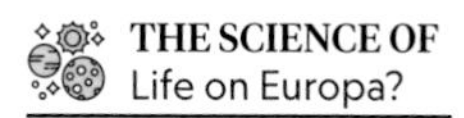

THE SCIENCE OF Life on Europa?

Data collected by the Galileo orbiter, launched from the space shuttle *Atlantis* in 1989, has raised questions about the possibility of life on Europa. Its surface is a shell of ice, roughly 1 to 10 miles (1.6 to 16 km) thick, covering a sea of briny water perhaps 60 miles (97 km) deep. Tidal tugs from Jupiter and neighboring moons Io and Ganymede keep Europa's interior ocean warm, and it has an atmosphere—very thin, but with molecular oxygen sputtering out from the icy surface. With water, minerals, and chemical energy, Europa's ocean contains all the ingredients necessary for life as we know it.

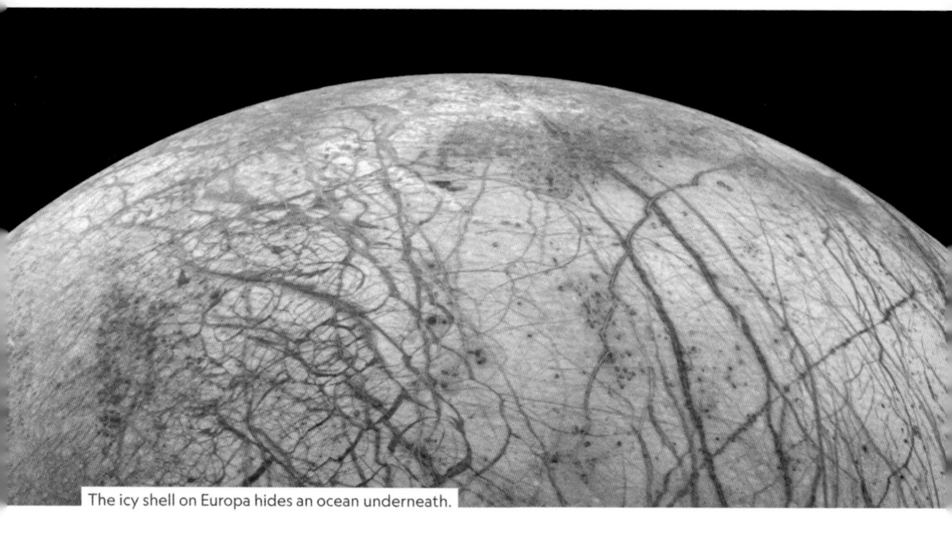

The icy shell on Europa hides an ocean underneath.

SATURN

SYMBOL: ♄

RADIUS: 36,184 mi (58,232 km)

MASS: 95.16 x Earth's

DISTANCE FROM SUN: 886,526,100 mi (1,426,725,400 km)

LENGTH OF YEAR: 29.5 Earth years

LENGTH OF DAY: 10.7 Earth hours

SATELLITES: 62

MAGNITUDE IN SKY: 0.6 to 1.5

FURTHER

Jupiter has its Great Red Spot; Saturn has its Great White Spot. The massive storm appears on Saturn's surface roughly every 30 years.

THE FARTHEST OF THE PLANETS seen by ancient astronomers, and easily visible to the naked eye, Saturn creaks along in a 29.5-Earth-year orbit that led astronomers in ancient Mesopotamia to dub it the "old sheep" of the sky.

But now Saturn is one of the solar system's standouts. While smaller than Jupiter, the planet's mass is 95 times that of Earth and its disk is 10 times wider. A Saturnian day is just over 10 hours, an extremely rapid rotation that has noticeably flattened the planet's disk at the poles. Meanwhile, the planet's razor-thin ice particle rings—170,000 miles (274,000 km) side to side but in spots only a few dozen feet thick—are one of the first targets an amateur astronomer is likely to hunt for.

On the Planet

To an observer, Saturn's atmosphere is all you can see. Its thick soup of gases—mostly hydrogen and helium—surround a body of liquid hydrogen and a core of rock and ice. Its cloud top is covered in a chilly fog of ice crystals. East-west winds create distinct bands like those on Jupiter, though the colors are more subdued shades of buff and white. However, images of Saturn in near-infrared light

Saturn in natural light and (above) infrared light

reveal dazzling contrasts and colorscapes, which scientists use to indicate the wavelengths of light invisible to the human eye. NASA's Cassini spacecraft spied a monster cyclone raging around the north pole with an eye that stretched 20 times larger than those seen on Earth.

Viewing Saturn

Almanacs and sky charts will provide information about Saturn's location and its rising and setting times. It will linger within a constellation for more than two years at a time. While visible to the naked eye as a bright-yellow-tinged star, a telescope is a necessity to view the planet in all its glory. Saturn at opposition is less than half of Jupiter's apparent size in the sky.

Saturn's rings cannot be distinguished through binoculars—a frustrating fact. But a small telescope will reveal them, while 6 inches (150 mm) of aperture will bring into view three moons and some surface details. Keep in mind that the appearance of the rings does change as Earth and Saturn orbit the sun. In between times of opposition, the globe casts a shadow on the near or far side of the rings.

Saturn is tilted to the ecliptic, like Earth, and when tilted toward us, we see a broad "top-down" view of the rings. The other side comes into view when the planet is tilted away from us. But roughly every 14 years, Saturn's rings are tilted edgewise toward Earth, making them all but disappear from view—an event that will occur in 2024 or 2025.

THE SCIENCE OF Saturn's Ears

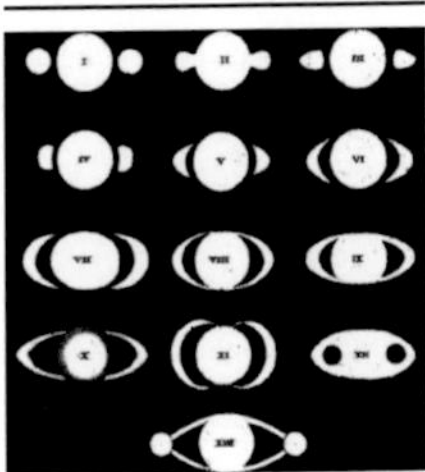

When Galileo first viewed Saturn in 1610, he saw bulges at its sides that looked like moons. Two years later they disappeared, only to reappear four years later as large ellipses. These changing observations were sketched in Galileo's notebooks, but he never solved the mystery. Some 40 years later, thanks to a more powerful telescope, Christiaan Huygens did: There was a ring around Saturn, tilted toward the ecliptic, that changed perspective and appearance through its orbit as the angle of observation changed.

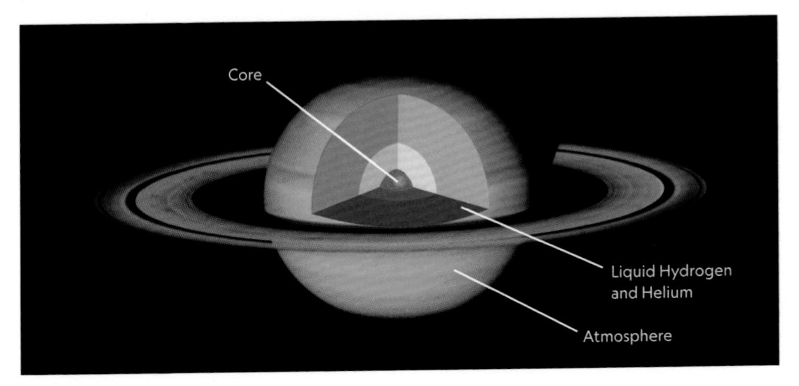

SATURN'S RINGS & MOONS

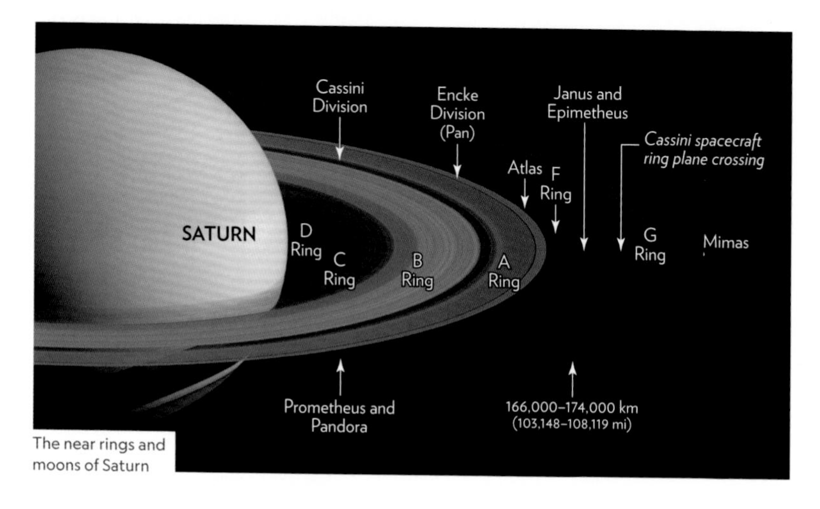

The near rings and moons of Saturn

FROM THE PIONEER 11 FLYBY in 1979 to the Cassini orbiter from 2004 to 2017, up close studies of Saturn have revealed much. The ice particles that compose Saturn's rings are as fine as the crystalline mist felt in a chilly fog and as large as a house.

NASA's Cassini also revealed that the rings are made up of thousands of bright narrow ringlets looking much like concentric grooves on an old phonograph record. Cassini actually zipped through the ring plane (between the F and G rings) and found that the Cassini Division contained relatively more dirt than ice. The material appeared similar to what was detected on the surface of Phoebe, one of Saturn's moons, lending credence to the idea that the rings are the pulverized remains of erstwhile moons, or comets and asteroids, that strayed too close.

FURTHER

Between Saturn's rings are multiple dark gaps of various widths. Backyard telescope observers can easily survey the big gap between its rings, known as the Cassini Division, that distinguishes Saturn's outermost A ring from the broad, bright B ring. Both rings brighten at the border of the Cassini Division. The fainter C ring follows B and is harder to see. More diffuse rings have been detected in the main ring structure: D, E, F, and G.

Saturnian Moons

There are six Saturnian moons visible through a small telescope: Titan, Enceladus, Tethys, Dione, Rhea, Iapetus. Its biggest and brightest, Titan, is even larger than Earth's moon, and the only one in the solar system sporting its own atmosphere. It's covered in liquid hydrocarbons and has a very Earthlike surface, including a scattering of liquid methane lakes and seas. Some scientists even think it could be capable of hosting extraterrestrial life.

In total, Saturn is also graced with at least 62 major moons of all sizes and compositions. Nestled within the gaps between Saturn's rings are the shepherding moons, a type of tiny moon that effectively herds the particles within the rings and keeps them in line. Shepherding moons run just along the inner and outer edge of the rings, creating a gravitational pull and tug that locks the particles in their relative positions, resulting in rings with sharply defined edges.

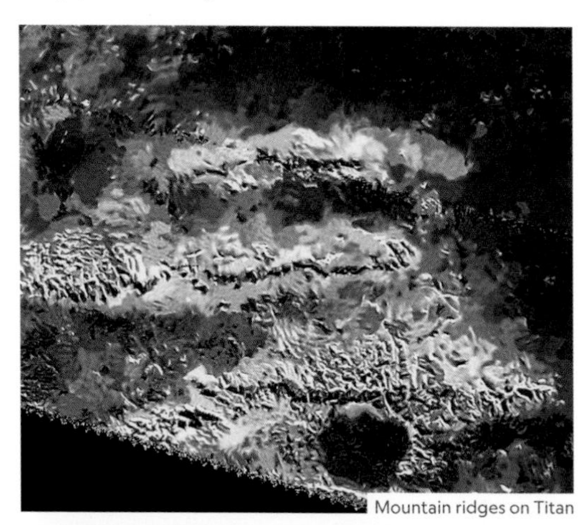
Mountain ridges on Titan

THE SCIENCE OF
Enceladus

Saturn's sixth largest moon, Enceladus, has become one of the most scientifically talked about planetary bodies thanks to its geologic activity. This 300-mile-wide (500 km) moon has its share of ancient cratered regions, covered in ice, that make it the most reflective body in the solar system. But they're mixed in with newer, much smoother terrain, filled with fresh fissures and more than 70 gushing geysers. These geysers send icy plumes of salt water and organic chemicals jetting into space at speeds of 800 miles per hour (1,300 km/h), continuously spouting onto the moon's surface. This feature offers convincing evidence that a reservoir of water lies beneath.

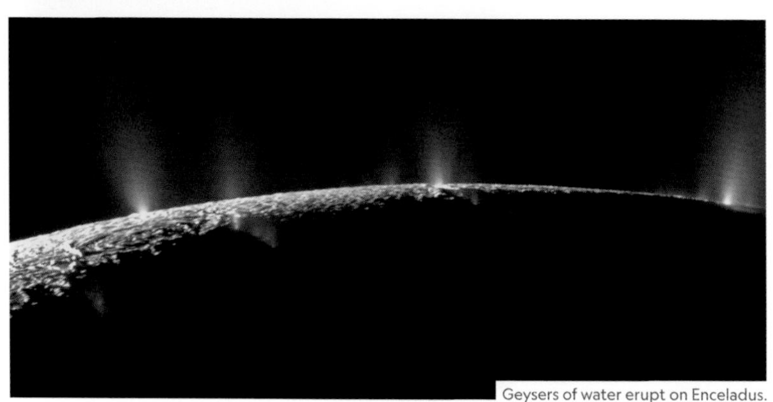
Geysers of water erupt on Enceladus.

URANUS & NEPTUNE

URANUS

SYMBOL: ⛢

RADIUS: 15,759 mi (25,362 km)

MASS: 14.54 x Earth's

DISTANCE FROM SUN: 1,783,744,300 mi (2,870,658,186 km)

LENGTH OF YEAR: 84.02 Earth years

LENGTH OF DAY: 17.24 Earth hours

SATELLITES: 27 moons

MAGNITUDE IN SKY: 5.3

FOR ANYONE training his or her telescope on these two distant planets, be prepared for a visually subtle outcome. Uranus and Neptune are so far away they will appear as relatively dim disks in your eyepiece—you'll be able to make out the planets' dominant color, but little in the way of details. But they make an interesting study simply because they allow us a glimpse at the edge of the solar system, where reflected sunlight takes more than three hours to reach your eye. Locating them requires star charts—they'll be essential—and using your scope. The fastest way to find both planets is to use a planetarium app or a GoTo robotic telescope.

Tilted Uranus

Uranus lies almost on its side, tilted 98 degrees to its orbital plane, the possible product of a massive collision. Because of its tilt, each of its poles gets about 42 years of sunlight before dropping into darkness. The planet is technically at the outer limit for the unaided eye, but as a practical matter binoculars will be needed. The planet is easily recognized as a small, uniquely green disk, a color imparted by traces of methane in its atmosphere. Through the eyepiece, it's about a 10th the size of Jupiter. Of Uranus's 27 moons, its largest—Titania and Oberon—can be spotted with a 6-inch (150 mm) telescope.

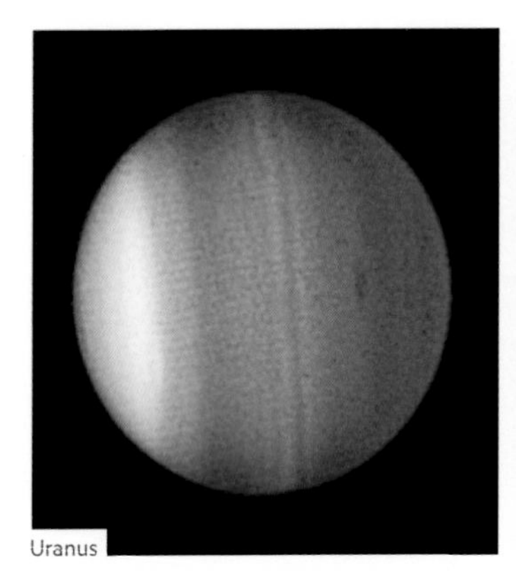

Uranus

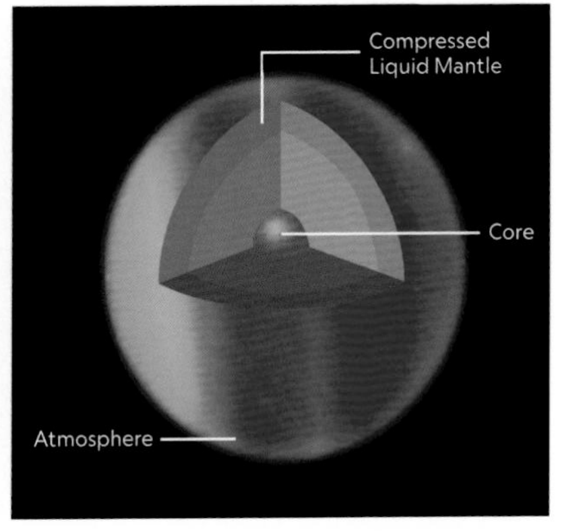

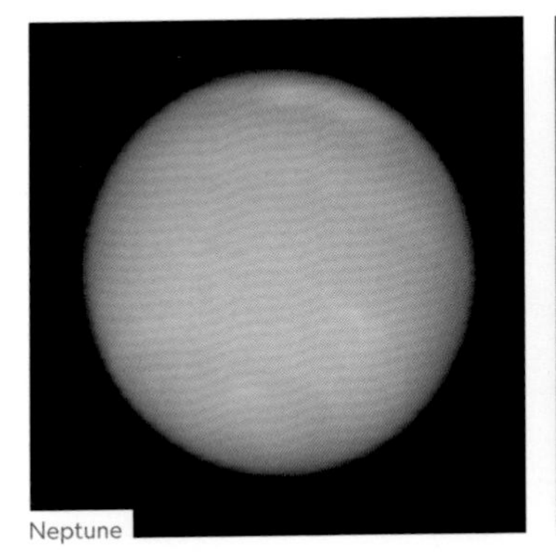
Neptune

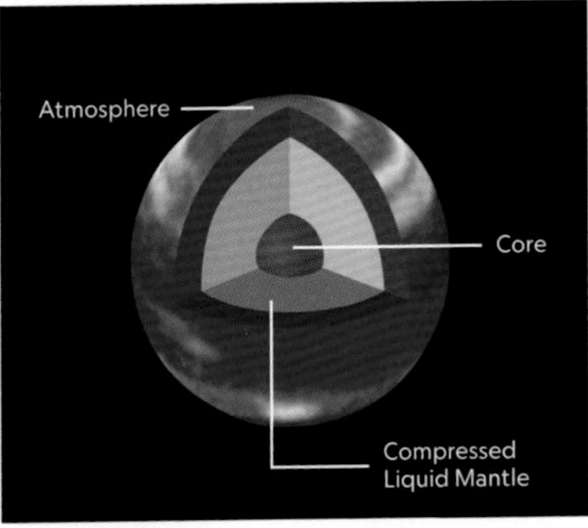

Discovery of Neptune

The astronomers of antiquity did not know about either Uranus or Neptune, though Galileo came close in the 1600s. He actually laid eyes on Neptune but regarded it as a star because his equipment was not powerful enough to discern the distant planet as a disk. And despite his intense exploration of Jupiter in 1609, his telescope's narrow field of view meant that Uranus—hovering just two degrees from the giant planet at the time—went unnoticed. It wasn't until 1781 that William Herschel made the definitive observations of Neptune.

Seeing Blue

The ice giant Neptune is as blue as the ocean, and aptly named after the Roman god of the sea. It is now thought to possess a massive ocean around its core of rock, all beneath an atmosphere thick with hydrogen and helium. Its bluish tint comes from clouds of icy methane that wrap around the planet.

In 1989 Voyager 2 detected faint rings around Neptune, uncovered a storm system dubbed the Great Dark Spot, and brought the number of known moons to 14. This outermost planet will require a finder chart, a telescope, and some patience to distinguish it from the stars around it. Look out for a gray-green disk about two-thirds the size of Uranus. Spotting the large moon Triton would require a telescope with at least an 8-inch (200 mm) aperture.

NEPTUNE

SYMBOL: ♆

RADIUS: 15,299 mi (24,622 km)

MASS: 17.15 x Earth's

DISTANCE FROM SUN: 2,795,084,800 mi (4,498,252,900 km)

LENGTH OF YEAR: 164.79 Earth years

LENGTH OF DAY: 16.11 Earth hours

SATELLITES: 13 moons

MAGNITUDE IN SKY: 7.8

FURTHER

One of Neptune's moons, Triton, is the only large satellite with a retrograde orbit, moving opposite to the planet's direction of travel. It most likely formed separately but was trapped by Neptune's gravitational pull.

PLUTO & BEYOND

SYMBOL: ♇

RADIUS: 715 mi (1,151 km)

MASS: 0.002 x Earth's

DISTANCE FROM SUN:
3,670,092,055 mi
(5,906,440,628 km)

LENGTH OF YEAR:
248 Earth years

LENGTH OF DAY:
6.387 Earth days

SATELLITES: 3

MAGNITUDE IN SKY: 13.6

Pluto (right) and its moon Charon

HOW ARE PLANETS DISTINGUISHED from the other objects orbiting the sun? Controversy over Pluto, a planet until 2006, helped crystallize the question. Spotting Pluto requires at least an 8-inch (200 mm) telescope, as well as the patience to watch for its subtle movement over several nights. Even sky charts will only provide you with a starting point. Pluto is so small and distant—nearly 39 AUs beyond Earth—that it will appear no different from the stars.

Pluto's Status

Pluto's stellar neighborhood contains hundreds of objects left over from the solar system's formation. In 2003, an object even larger than Pluto—and with its own moon—was discovered, forcing a decision that changed Pluto's status and that of objects like it. In debating Pluto's fate, the IAU cited three tests for planetary status: First, the body must orbit the sun; second, it must have a close-to-round shape created by its own gravity; and third, it must have cleared the neighborhood of debris along its orbit. Pluto failed the third test, as it circles in the company of many other objects beyond Neptune.

Pluto thus became the first in a new category: dwarf planet. The second dwarf planet discovered is Eris. The two were also placed in a special category of dwarf planets, known as plutoids because they orbit beyond Neptune—a name that honors Pluto's former status.

Into the Kuiper Belt

Pluto resides in the Kuiper belt, a debris field that extends from about 35 to 55 AU, and outside that lies a much wider scattered disk of small, icy bodies. The Kuiper belt is an area where planet forming appears to have stopped. Objects are too spread out and orbiting too slowly to bunch, collide, and coalesce into larger bodies. They remain slow-circling planetesimals or protocomets that occasionally dislodge and travel through the solar system.

When NASA's New Horizons probe flew by Pluto and its moon Charon in 2015, it discovered that both are amazingly diverse, complex worlds with dramatic canyons, mountain ranges, and smooth plains that show evidence of recent resurfacing.

FURTHER

The list of dwarf planets is likely to grow in the coming years, perhaps by the hundreds. As of spring 2018, the IAU had acknowledged five. Pluto, Eris, Makemake, and Haumea are all plutoids that circle beyond Neptune. The fifth, Ceres, is located in the asteroid belt.

Large debris in the Kuiper belt

PHOTOGRAPHING THE NIGHT SKY

EVERYTHING YOU CAN SEE in the night sky—and many objects too faint to see—can be photographed. While astrophotography can employ complex computer-guided telescopes and specialized cameras, some of the most compelling images of the night sky require no more than some off-the-shelf camera gear.

Sky Subjects

The dark of night presents an inexhaustible array of subjects for any photographer. Landscapes that look conventional by day take on an exotic character when photographed by night, using nothing but moonlight and starlight for illumination. Add a colorful display of auroras and you have a unique, award-winning image. Many nightscapes can be captured with exposures no longer than 30 seconds. Extend the exposure time to minutes, or even an hour, and it will capture the colorful trails left by stars as they wheel about the heavens.

Placing your camera on a telescope or motorized mount that can track the turning sky ensures you don't get undesired star trail lines during the long exposure time required for the incoming light to

Capturing the northern lights requires long exposure time.

build up on the camera's sensor. The image records nebulae and star fields too faint for the eye to see, revealing a universe beyond the threshold of human vision.

The Best Cameras

Digital cameras have all but completely replaced film in astrophotography. They offer instant results, an essential aid to nighttime imaging when exposures must often be guessed at and the subjects can be too faint to frame in the viewfinder. Digital cameras are also far more sensitive when shooting dark sky scenes, picking up in seconds or minutes what film might have taken hours to record.

A glance at the specifications of digital cameras might suggest that sheer megapixel count determines image quality. Not so. For long-exposure images of faint nighttime subjects, it's noise—not a lack of pixels—that distracts from the beauty and ruins astrophotographs. Electronic noise is present in all digital cameras, and in long exposures it can build up to pepper an image with colored, grainy specks.

The cameras best able to capture clean, noiseless images of the night sky are digital single-lens reflex (DSLR) cameras—the kind that allow users to change out lenses and have an optical viewfinder that allows the photographer to compose the shot while looking through the lens. DSLRs have larger digital chips with larger individual light-sensitive pixels than do smartphones or small, pocket-size cameras. The larger chips are able to record more light in a given exposure, yielding images with less unwanted noise and more wanted signal.

A rock-steady tripod is essential.

Basics for Shooting the Sky

For best results when shooting long-exposures with DSLRs, keep the following tips in mind. Turn on any long-exposure noise reduction setting, which can usually be found under Custom Functions or on the Shooting menu. Switching on High ISO Noise Reduction can also help. Also, look for DSLRs with manual controls and the possibility of using wide-angle lenses and a wide aperture. This allows for faster exposures, offering better quality and lower noise levels.

When taking night sky images, the camera has to be held rock steady. That demands a sturdy tripod. Another essential is a remote release or self-timer, which allows the shutter to be triggered without jiggling the camera and can be held open for as long as desired. Extra batteries are also a good idea, as long exposures on cool nights can drain camera batteries quickly.

SIMPLE TECHNIQUES

Open the camera shutter for five to 30 minutes to record the movement of stars.

SOME OF THE MOST DRAMATIC astronomical photographs are taken with no more than a standard camera on a tripod. What makes for a great nighttime photo is the same that makes any photo a winner—good composition and an eye-catching subject.

Nightfall

After nightfall, moonlight can provide enough illumination to light the ground and reveal details. An exposure of 20 to 40 seconds (at f/2.8 and ISO 400) produces a scene that looks like daylight, complete with blue sky, but the sky is filled with stars. But to capture deep-sky Milky Way arches, dark and moonless nights are required; make sure to plan photos around moon phases to ensure the darkest possible skies. To add some Earthly scenery, a flashlight can be used to "paint" nearby objects—trees, tents, houses, and even people—while taking a long exposure.

Constellations

The beauty of digital cameras, particularly DSLRs, is that they can pick up lots of stars in a relatively short exposure time

Open the lens to about f/2.8 (or with slower "kit" zoom lenses to wide open at f/3.5 to f/4) and set the camera anywhere from ISO 400 to ISO 1600 (for greater sensitivity). Then frame a constellation and open the shutter for 20 to 40 seconds. This is best done from a dark rural site, but even from the city this technique will record bright stars and constellation patterns, though exposures may have to be limited to a few seconds. A normal or wide-angle lens is best for framing most constellations. From a dark site, exposures no more than a minute long at ISO 1600 can pick up the glowing star clouds of the Milky Way. When the image pops up, you'll be amazed at the vivid colors and detail.

Star Trails

As Earth turns, the entire sky appears to rotate about the celestial pole. In the Northern Hemisphere, this point lies near Polaris, the North Star. Open the shutter for five to 30 minutes and the image will record the stars as streaks circling the polestar.

The trick is to take the lens down to between f/4 and f/8 and reduce sensitivity to ISO 100. How long the camera should be left depends on its quality, the sky's level of darkness, and the temperature. Low temperatures can keep camera sensors cold enough to reduce camera noise that might otherwise build up and allow for much longer, quality exposures—even up to one or two hours.

Experiment with camera angles and use foreground objects to give your images some context. You should also be prepared to snap dozens, if not hundreds, of photos in a single session.

A light-sensitive camera can detect the variation of color among the stars in Taurus.

EQUIPMENT

JUST ABOUT EVERYONE who owns a camera and a telescope wants to connect the two. But long exposures of nebulae and galaxies require complex equipment and techniques best left for when a photographer has gained experience with simpler methods and is willing to pay for the gear required. Try these techniques for your first few outings, whether you're shooting with a smartphone or something far more sophisticated.

Camera-to-telescope adapter

Camera and telescope attached

Snapping the Sky

Using nothing more than a simple smartphone, you can capture surprisingly detailed wide-angle photographs of the brightest stars, constellations, and planets, but also the fainter Milky Way, meteors, and aurora. Make sure you have a sturdy mount, avoid using zoom, lock the focus, and boost the brightness in order to really bring out the stars. Download a night photography app that optimizes the phone's camera for low-light conditions and opens the camera shutter for long exposures.

Adapting a Telescope

Any binoculars or telescope can be turned into a super-telephoto lens for dramatic close-ups of the moon and planets. An easy way to do this is to employ your smartphone by placing it in front of the viewfinder or eyepiece using an adapter. Or remove the lens from a DSLR camera and replace it with a camera-to-telescope adapter tube, equipped with a T-ring that allows the adapter to click into the camera's lens mount, just like a lens. The required ring and adapter are available at any telescope dealer.

With its adapter in place, the camera then slides into the telescope's focuser in place of the eyepiece. Focusing requires care, as the image through the camera viewfinder can be dim. Focus so that the moon's limb, or the edges of shadowed craters, look as sharp as possible.

Exposures are short, which isn't surprising, since the moon is a bright, sunlit rock. The precise exposure time will depend on the telescope and moon phase but are typically 1/15 to 1/500 of a second at ISO 100 to 400. These are short enough that an electric drive on the telescope is not essential, but it's recommended to keep the moon framed during a shooting session.

Motorized Shots

A step up in complexity is sky-tracking photography, where the camera tracks the sky during a long exposure using a motor that compensates for Earth's rotation. The resulting images can be spectacular, revealing night sky objects in a whole new light. Popular time-lapse "nightscape" videos are captured using variations of this method. In this scenario the camera, equipped with a wide-angle, normal, or telephoto lens, either directly attaches to a motorized mount or rides "piggyback" on the side of a telescope. The camera rides astride the scope, rather than looking through the eyepiece. The telescope itself must have a tracking motor, and it must have an equatorial mount aligned to rotate around the celestial pole. The telescope's manual should contain instructions on how to perform the required polar alignment. The proper alignment is necessary for achieving the pictures you want.

Smartphone mounted with a spotting scope

While the setup is more demanding, the results can be spectacular. Exposures of two to four minutes at f/2.8 and ISO 800 to 1600 reveal countless stars and faint nebulae in vibrant, rich color, while the stars remain pinpoints because the telescope's tracking system counteracts the streaking effect of Earth's rotation.

A piggyback mount on a telescope with a tracking motor

CHAPTER 6
COMETS & METEORS

COMETS

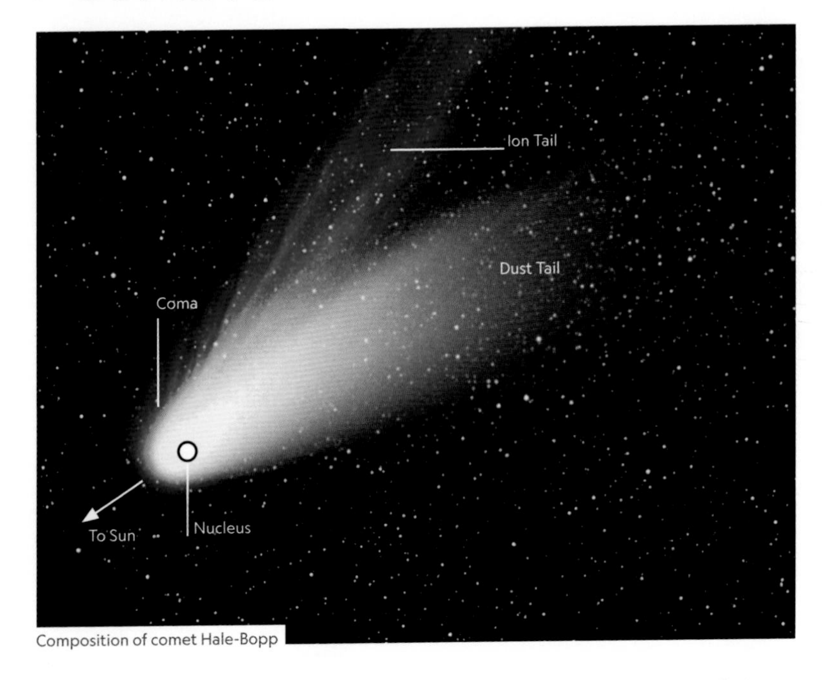

Composition of comet Hale-Bopp

AS THE flat plane of the solar system took shape, billions of balls of frozen, dust-laden gas formed. Many of these cosmic snowballs are thought to have been swept by the gravitational forces of Jupiter and Saturn out to the distant Oort cloud. Others formed just outside the orbit of Neptune, creating what is now known as the Kuiper belt and a far-flung, overlapping area called the scattered disk. We see these ancient objects when they are nudged into a new orbit and soar through the inner solar system as comets. Though not as constant, comets are much more numerous than the planets.

FURTHER

King Louis XV dubbed French astronomer Charles Messier the "comet ferret" for the way he was constantly discovering new ones.

Composition

A stew of frozen chemical compounds, comets begin to warm up as they approach the sun and release the gases stored during the solar system's early days. The gases form a glowing head, or coma, around the frozen nucleus, and the solar wind shapes a flow of charged ions into a glowing gas tail. Meanwhile a stream of dust, also shed by the nucleus, forms a second tail.

Once in an orbit that will pass through the inner solar system, comets tend to follow an elliptical path. Those dislodged from the Oort cloud become long-period comets or get slung from the solar system, while those that originate in the Kuiper belt or scattered disk orbit in a shorter period. Comets typically are large enough to survive perhaps hundreds of trips through the inner solar system before disintegrating.

Observation

You can observe predicted comets or hunt for new ones. About two dozen are discovered each year, a few bright enough to be spotted by amateur equipment—especially near the sun, where robotic, professional surveys can't observe. If you see something not on a chart, make sure it is not a star, look for a tail, and then recheck it to see if it has moved. Make notes on its location and magnitude, sketch its appearance, and have someone confirm what you've seen. Also, make sure it's not a known object. If it seems like a comet you can notify the Central Bureau for Astronomical Telegrams in Cambridge, Massachusetts.

THE SCIENCE OF Organic Soup

After a decade-long trip, the European Space Agency's Rosetta spacecraft reached comet 67P/Churyumov-Gerasimenko in 2014. It orbited for two years, sending robotic lander Philae to the comet's surface to study its topography and composition. According to the data it collected, organic molecules—carbon-based ones like proteins, carbohydrates, and nucleic acids—make up about half of the dust that 67P emits. Scientists describe these chemicals as frozen primordial soup, supporting the theory that comets may have seeded the early Earth with the ingredients for life.

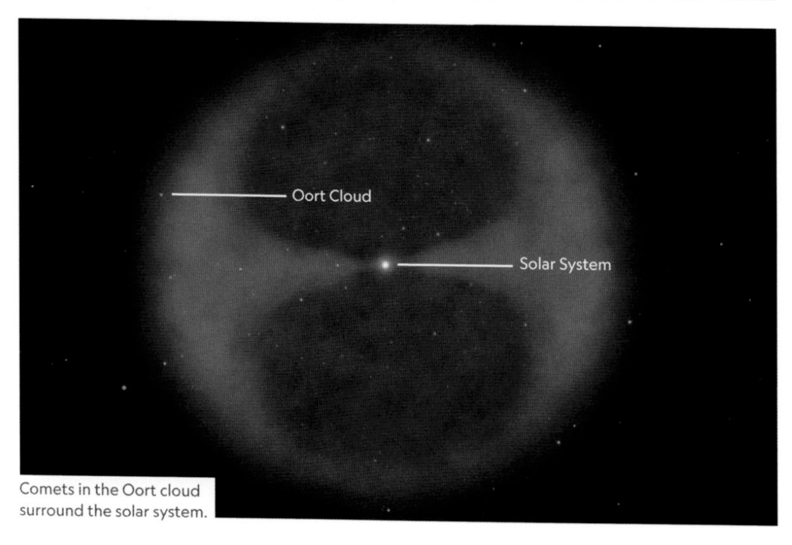

Comets in the Oort cloud surround the solar system.

FAMOUS VISITORS

IN THE LATE 1600s, with Isaac Newton's math-based view of the world just beginning to take hold, astronomer Edmond Halley tried to use new ideas about gravity and motion to see if he could predict—as Newton's theories seemed to allow—the return of any of the comets noted in historical records. Establishing tables of comet observations, Halley noted that one of these wandering bodies had turned up in 1531, 1607, and 1682—and it was headed in roughly the same direction each time. He ventured a bold guess: The three comets were the same object, and it would appear again in late 1758—a prediction that proved to be accurate.

Comet Sightings

In 1973, comet Kohoutek was a widely publicized bust that—despite the fact that it was visible to the naked eye—did not seem to match the "comet of the century" billing it received when discovered. The truly bright ones don't happen very often, but scientists and sky-watchers have still had plenty to work with. Three bright comets put on a show in the 1990s. Comet Hyakutake in 1996 and comet Hale-Bopp in 1997 were watched by professionals and amateurs alike, but both are long-period and won't be back soon.

Red-tinted comet Siding Spring passes between two bright stars in a time-lapse image.

FURTHER

The record for the largest comet nucleus goes to comet Hale-Bopp, measuring more than 60 miles across (97 km), while comet McNaught had a gas tail measuring more than 140 million miles (225 million km) long.

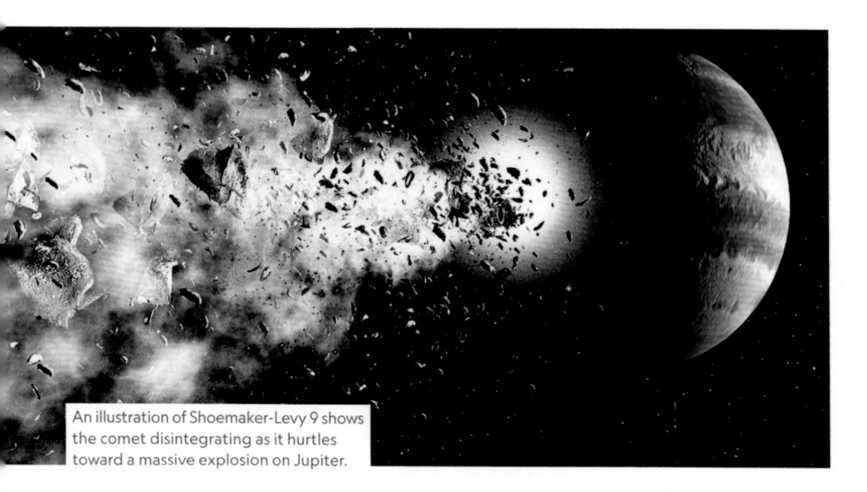

An illustration of Shoemaker-Levy 9 shows the comet disintegrating as it hurtles toward a massive explosion on Jupiter.

In 1994, comet Shoemaker-Levy 9 earned special attention when it was captured and ripped apart by Jupiter's gravity. The collision of its shredded pieces with the massive planet caused a series of spectacular explosions. The encounter and its aftermath were studied closely for clues about the effects of such interplanetary calamities. Comet McNaught—the Great Comet of 2007—was the brightest visitor in many decades, but it is on an apparent path to leave the solar system and never return.

Missions

There have been robotic encounters with 12 comets, conducted by the United States, the European Space Agency (ESA), Japan, and the old Soviet Union. The first spacecraft to visit a comet, NASA's ICE probe, was back in 1978, and it studied comet Giacobini-Zinner. The mission showed evidence that comets were made of a mixture of ice and rock, proving the "dirty snowball" theory. In 1986 Japan sent two probes, and ESA sent one of its own, to study famous comet Halley. The first comet dust samples delivered to Earth were collected and delivered by NASA's Stardust probe after its visit to comet Wild 2 in 2004. The first mission to study a comet's interior was NASA's Deep Impact in 2005, when a probe was propelled into comet Tempel 1 to excavate a crater for analysis.

THE STORY OF Ancient Views

Halley's comet has been a regular visitor for thousands of years. Its first predicted reappearance happened in 1758, but confirmed sightings go back as far as ancient Chinese astronomers in 240 B.C. New studies indicate there was a large meteor fall in ancient Greece in 466 B.C., which coincides with when Halley would have flown past Earth. Calculations and computer modeling show that the comet's tail and its debris would have swept directly across Earth. This encounter would have made the tail appear very large, with a flurry of shooting stars from the onslaught of comet particles slamming into our atmosphere.

METEORS

A meteor falls during the Leonid shower.

METEORS, OR shooting stars, are a fairly easy target for a patient observer. Under dark skies on any average night, a sky-watcher can expect to see one every 15 minutes or so. As Earth progresses in its orbit, the planet encounters upward of half a billion of these objects every year.

Meteoroids

Meteors begin as meteoroids—interstellar dust and debris. Individual particles are typically small, though some may be several feet wide. At that larger scale, the distinction between meteoroids and asteroids is somewhat arbitrary (the International Astronomical Union defines meteoroids as "considerably smaller than an asteroid and considerably larger than an atom or molecule"). Larger ones are thought to be pieces of asteroids, planets, or our moon that have broken off and gone adrift. But most meteoroids are small grains of interstellar material, often the "ash" burned off the outer layers of comets and left in massive trails around the solar system. The most prevalent are made mostly of silicates and other

FURTHER

While many of the meteors we watch rain down are sand-grain size, the vast majority are no bigger than a mote of dust—a fraction of the width of a human hair. And while larger ones ionize in the upper atmosphere, these micrometeors continually float to Earth's surface day and night—as much as 100 tons (90 t) of the stuff daily—across the entire planet. That means micrometeoritic dust covers everything, and you can collect it for further study. Most meteorites contain a high iron content, which means they stick to magnets. Run magnets near the downspouts of your rain gutters and you're likely to collect two or three tiny extraterrestrial particles per several square feet. Look though a microscope to examine your treasures.

rocky material; some appear to have been formed from the same early solar system matter that produced the sun. About 6 percent are iron meteorites, composed of iron and nickel. A small number of stony-iron meteorites are a roughly equal mixture of the two.

Earthly Contact

These small bodies become shooting stars when they enter Earth's atmosphere. Under its gravitational pull they reach speeds of anywhere from 20,000 to more than 160,000 miles per hour (32,000 to 257,000 km/h). Friction from the atmosphere heats them to about 3000°F (1650°C), creating the glowing streaks that inspired the term "shooting star." Under such extreme heat and pressure, most meteors vaporize before they get within 50 miles (80 km) or so of Earth. Most are so small that they last barely a second before being destroyed, and they are sometimes so quick and dim that we can't even see them. Called bolides, or fireballs, these unusually bright meteors create long, bright, broken streaks with smoky trails, and sometimes even a sonic boom. Because sound travels more slowly than light, wait five minutes after the visual explosion to hear the boom.

Occasionally, pieces of meteors reach Earth as meteorites. Most are harmless, but the biggest ones have caused mass extinctions, like the one likely to blame for ending the reign of the dinosaurs.

FURTHER

Sixty-four or so of the meteorites discovered on Earth are known to be pieces of the moon, while 34 are fragments of Mars. There may well also be meteorites from Venus, Mercury, and other bodies in the solar system, but none has yet been specifically identified.

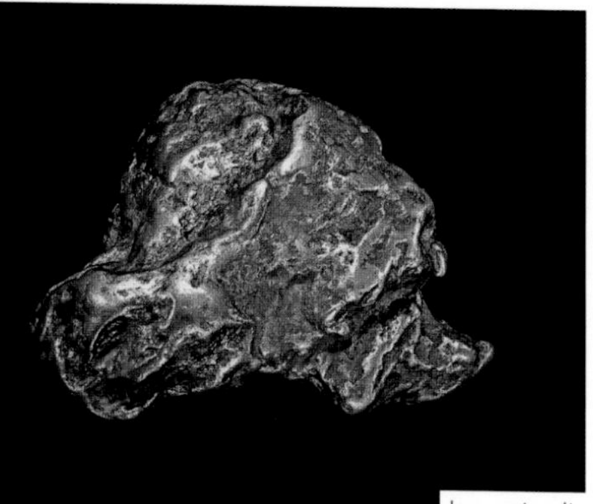

Iron meteorite

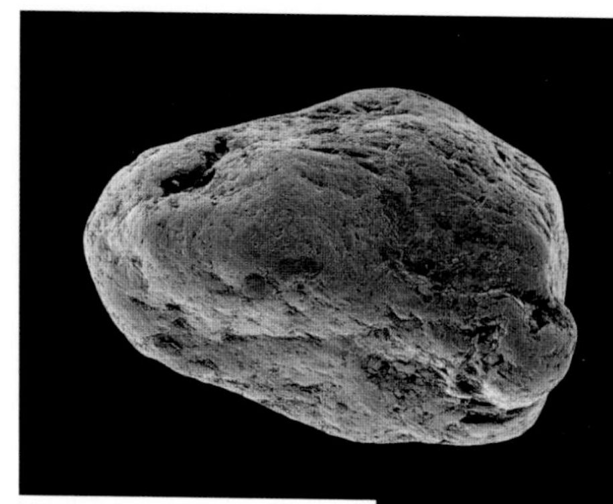

Micrometeorite under magnification

FALLING TO EARTH

THE FACE OF THE MOON tells the story of the solar system's violent early years. Our planet has been bombarded too, but whereas the moon's stable geology has maintained evidence of its contact with asteroids and meteoroids, Earth's tectonic plates, vast sea cover, and changing surface have hidden or erased most of its dents and dings.

THE STORY OF
Aboriginal Iron Tools

In 1897, explorer Robert E. Peary deduced the source of the iron used in tools among Greenland Inuit when a local guide helped him track down a meteor split into three deposits. With great difficulty, Peary brought a piece weighing 34 tons (30.8 t) back to New York. Decades later, his wife, Josephine, sold the meteor to the American Museum of Natural History in order to send a ship to retrieve her husband from Greenland.

Crashes & Craters

The surface of Earth bears the scars of its encounters with interstellar agents: 175 craters and larger impact basins worldwide, all associated with comets, asteroids, and meteoroids. Some, such as the one in that struck Sudbury, Canada, 1.8 billion years ago, contained enough metal to support local industries and enrich the surrounding farmland. With a width of three-fourths of a mile (1.2 km) and a depth of 600 feet (183 m), the 50,000-year-old Barringer Meteor Crater in Arizona resembles the lunar surface.

Found on Mexico's Yucatán Peninsula in the 1970s, the 110-mile (177 km) Chicxulub crater has been buried by sediment, but exploration has revealed it bears the geologic markings of an asteroid strike—glassy tektite rocks

A 50,000-year-old meteor strike shaped the Barringer Crater in Arizona.

THE SCIENCE OF
The Siberian Impact

Blasted trees in Tunguska, Siberia

On June 30, 1908, a massive comet or asteroid visited remote Siberia near the Tunguska River. It detonated near the surface of Earth, so no impact crater was found, but crash-site investigations revealed that some 800 square miles (2,072 km^2) of forest had been flattened, with millions of trees splayed out in a radial pattern that marked the epicenter of the blast. Reports from the time of impact show seismic shock waves and bright skies registered as far away as England. While there is still some debate as to the identity of the impactor, it is generally accepted that it was a massive space that entered the atmosphere at extreme speeds. The resulting fireball caused the surrounding air to reach 44,500°F (24,704°C) and release energy equivalent to 185 Hiroshima bombs.

and large amounts of iridium, among other things. The object is estimated to have been 6 to 12 miles (10 to 19 km) wide and has been dated to the end of the Cretaceous period, which is consistent with the disappearance of many of Earth's species, suggesting that the asteroid's dust blocked sunlight, killed much of the plant life, and caused temperatures to plummet.

Impact Odds

On February 15, 2013, a 60-foot (18 m) asteroid entered the atmosphere undetected and exploded high above the Russian city of Chelyabinsk. The resulting blast wave blew out windows, damaged buildings, and injured some 1,500 people. Meteor impacts that could cause regional, or even global, catastrophe are blessedly rare. By some estimates, a collision large enough to cause a mass extinction will occur every 100 million years. Objects big enough to destroy a city are estimated to hit perhaps every thousand. Plus, we've gotten good at finding near-Earth objects: Surveys have identified more than 90 percent of near-Earth asteroids larger than half a mile (1 km) and tracking their movement. But we can only detect big rocks, at best, a few days out from the Earth-moon system.

METEOR SHOWERS

An interpretation of the November 1833 Leonid meteor shower

THE OCCASIONAL SHOOTING STAR that enlivens a campout or a stroll on the beach is called a sporadic meteor, and these can occur at any time of the night all year long. But about 30 times a year Earth passes through solar system debris and creates a meteor shower—or a spectacular (and rare) meteor storm.

Forecasting Showers

Since meteor showers occur regularly, their appearances are well publicized in astronomy magazines, on websites, by organizations like the International Meteor Organization, and probably by your local meteorologist as part of the weather report. There is no shortage of information about when annual showers are expected to occur and

FURTHER

The Leonid meteor storm on November 16, 1833, is estimated to have produced up to 200,000 shooting stars an hour!

peak. You can send reports of meteor events to data clearinghouses like the American Meteor Society, which use them to create statistical studies and possible alerts for the recovery of possible meteor debris.

Observation

A little preparation will make for more successful viewing. Choose an observation location in the countryside and time your observations for a less-bright moon phase. First and foremost, be conscious of where the particular shower will originate—its radiant—and position yourself facing away. Watch for meteors at least halfway up the sky.

Annual showers are typically named for the constellations from which they appear to emanate. If Leo is near the horizon in mid-November, when the annual Leonid ("children of Leo") shower occurs, you will need to find an adequate viewing spot. You may find out that a shower is peaking during daylight hours or under bright moonlit skies, giving you the choice of accepting a less-than-optimal view at night or early in the morning, or opting out until next year. Don't worry if you miss a shower's peak time; shower activity can last as much as a week before and after the peak night.

Meteors shoot across a time-lapse exposure of star trails.

FURTHER

The International Meteor Organization depends on information from amateurs to expand its database of observations, but it counts on observers to have enough skill to vet what they see before submitting it. For practice, take notes and try to capture as much detail as possible, including brightness, color, duration, and the length of any shooting star streaks across the sky. Identifying the radiant isn't difficult—individual meteors in a shower will track back to a common point. If possible, record the declination and right ascension from which the shower emanates. Try to record details about the sky condition. Take note, for example, of the faintest star you can see to set the limiting magnitude. You can estimate the speed at which the meteors are traveling by noting how many degrees of sky are covered in a second, and assess their brightness by learning to compare them with visible stars.

ANNUAL SHOWERS

ANNUAL METEOR SHOWERS can vary widely in their intensity, but a handful are among the most consistent, and all are worth a look for both serious hobbyists and casual naked eye observers.

Winter Showers

The Geminid shower is one of the densest annual showers, producing upward of 120 meteors an hour. The Geminids fall in mid-December and are one of the best to view in the evening (as opposed to after midnight), since the dust trail meets Earth's orbit opposite the sun. The source of the meteors is object 3200 Phaethon, an asteroid that may well be an "extinct" comet nucleus, one whose ices have become exhausted or deeply buried by surface dust.

FURTHER

The intensity of meteor showers is related to how soon after a comet's passage Earth intercepts its dust trail.

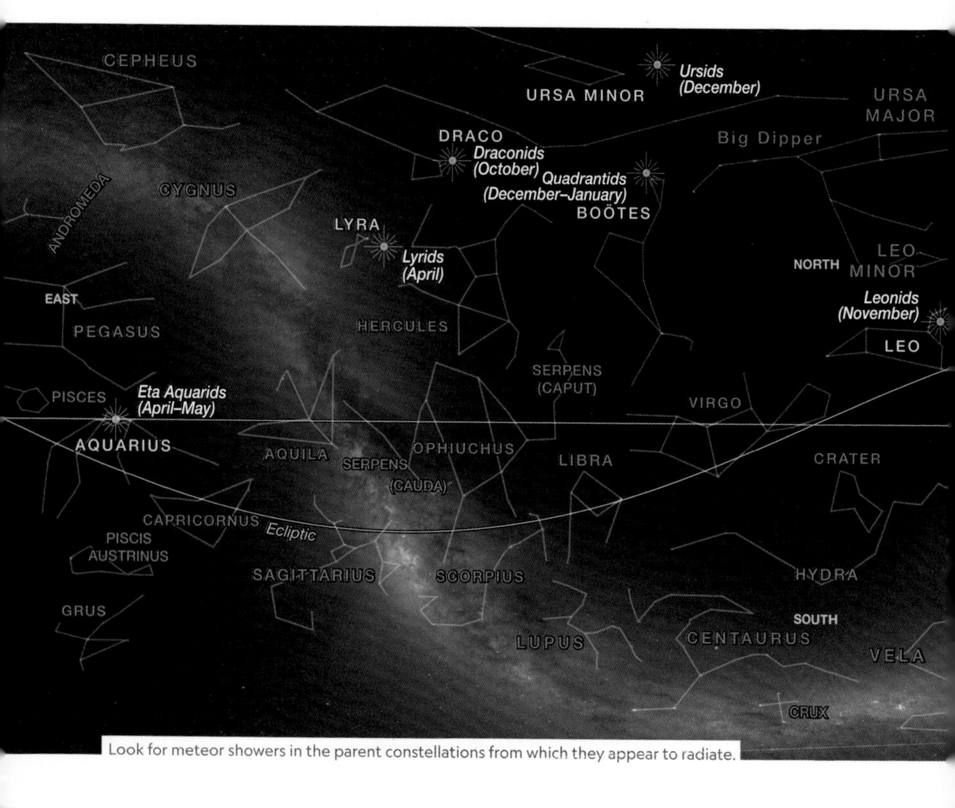

Look for meteor showers in the parent constellations from which they appear to radiate.

Spring & Summer Showers

Halley's comet left a dust trail in its wake that Earth approaches twice a year, producing two meteor showers: the Eta Aquarids in spring, which produce up to 20 meteors per hour at peak hours, and the famed Perseids in summer, recorded as far back as A.D. 36. Both are very reliable and swift, with rates of 80 to 120 per hour.

Autumn Showers

Emanating from Orion, the mid-fall Orionids peak at around 25 meteors an hour. The November Leonids, emanating from Leo, have been the source of repeated, intense meteor storms on a 33-year cycle, but they usually peak at only 15 to 20 meteors per hour. The Leonids are known to produce long-lasting trains across the sky.

MAJOR METEOR SHOWERS

QUADRANTIDS
DATES OF ACTIVITY: Jan. 1–5
PEAK: Jan. 3/4

ETA AQUARIDS
DATES OF ACTIVITY: Apr. 19–May 28
PEAK: May 4/5

PERSEIDS
DATES OF ACTIVITY: July 17–Aug. 24
PEAK: Aug. 12/13

ORIONIDS
DATES OF ACTIVITY: Oct. 2–Nov. 7
PEAK: Oct. 20/21

LEONIDS
DATES OF ACTIVITY: Nov. 6–30
PEAK: Nov. 16/17

GEMINIDS
DATES OF ACTIVITY: Dec. 4–17
PEAK: Dec. 13/14

A GUIDE TO BINOCULARS

TELESCOPES LIMIT YOU to one eye—an awkward situation for the brain. The amount of sky seen through the lens, called the field of view, is narrow, making it harder to find your target. Telescopes can be bulky and hard to set up too—a disincentive if the night is chilly or time is short. For ease of use, portability, and economy, all while still adding remarkably to the range of what you can see, a standard pair of binoculars is a great piece of equipment for the beginning sky-watcher.

Multicoated lenses will look blue or green.

Rubber coating is shock-absorbent and offers a comfortable grip.

Choosing Size

An important point to keep in mind when purchasing binoculars is that weight matters. You'll be holding them and looking up, so some of the heavier models may prove tiring as you scan the sky. You can, of course, invest in a tripod, but the ability to casually scan overhead without a lot of setup is one of the advantages of binoculars.

Choosing the right size, as measured by the level of magnification and the size of the objective lens (the outer lens), is crucial too. You'll likely see both stamped on the glasses somewhere: listed as 7 × 35 or 10 × 50, the first number represents the magnifying power and the second is the diameter of the lens in millimeters.

For general viewing a 7 × 50 configuration is the recommended choice. The 50-millimeter objective lens is large enough to gather the light needed for serious observation but compact enough to remain on a neck strap. The field of view is wide, and the image produced by a 7x magnification can be steadied by hand. Remember that binoculars also magnify the effects of trembling fingers: Even the jump to a 10 × 50 pair might require a tripod to control the jiggling. A great way o eliminate any shaking is to invest in mage-stabilized binoculars. Motion sensors and microprocessors control the shape of the prisms and compensate for any movement, offering seemingly rock-steady views. Just know these great views come at a higher price.

Buying Tips

The binoculars' prisms should be made of BaK-4 glass as opposed to BK-7. It will cost more, but is worth it for the extra brightness. You can tell by holding the glasses at roughly arm's length and examining the eyepieces. If the bright circle of light you see—the exit pupil—has any gray edges, the binoculars you are holding use BK-7 glass.

Shop for glasses with multiple-lens coatings. They're better at transmitting light (you can tell by shining a flashlight through the outer lens and tilting the glasses back and forth; coated glass will look blue or green). While in the store, test for alignment, also called collimation:

Large lenses offer advanced observation but will be heavy.

Focus on an object, then alternate closing your eyes—the image should stay stable. Hints of rainbow color around bright reflections or double images indicate lower quality optics.

Giant astronomy binoculars with lenses of 3 or 4 inches (80 or 100 mm) or more can have prices equal to, or even exceeding, those of telescopes. They are often too heavy to steady by hand and will require a sturdy tripod. For novices, this is not so much of a step toward a telescope as a substitute for one.

Binoculars with an image stabilizer offer steady views without a tripod.

CHOOSING TELESCOPES

INSTRUMENTS FOR AMATEUR skywatchers fall into three main types: refractors, reflectors, and catadioptric telescopes. Each one has its own advantages, and consideration should be given to how and when you will be using your gear. Even before looking at individual instruments, ask yourself a few questions: What kinds of objects do you want to observe? How much weight do you want to carry? What's your budget? Will you be observing mostly under light-polluted skies on your downtown apartment balcony, from your suburban backyard, or from a country cottage where you have pristine skies? While a larger aperture—the diameter of its main optical component—makes for better views, it also makes the instrument increasingly bulky and expensive. It's worth learning how to choose the right telescope for you. Telescopes are advertised by the size of the aperture of the main objective, so a 6-inch telescope has a main mirror or lens measuring 6 inches, or 150 millimeters across.

Types

Refractors look like what most people envision when you say the word "telescope." These feature an objective lens at the front end of the telescope tube, which collects light and directs it toward an eyepiece at the other end of the tube. The eyepiece then magnifies the image.

Reflectors do away with the objective lens and instead use a concave mirror at the bottom of the telescope tube. Incoming light falls down the tube onto the mirror and is reflected upward toward a secondary mirror, which directs

A 6-inch (150 mm) Dobsonian reflector is a high-value all-purpose telescope.

the light through the side of the tube to the eyepiece, which then magnifies the image.

Catadioptric telescopes combine a corrector lens with a mirror. Light enters the tube through the transparent corrector lens, bounces off a main mirror at the back of the tube, goes up to a secondary mirror, and reflects back down the tube where it passes through a central hole in the main mirror to the eyepiece, which lies behind.

Pros & Cons

Each type of telescope comes with pluses and minuses. Refractors, for example, tend to have a long focal length afforded by a longer optical tube. The higher number translates to a higher focal ratio that offers more sharply contrasted images, while lower focal ratios such as those you'll typically find in reflec tors, allow more sensitivity to light and wider fields of view.

If your main interests lie with the moon, the planets, double stars, and other such high-magnification phenomena, you might be best served with a refracting telescope. If, on the other hand, you are more interested in seeing star clusters, comets, and nebulae, a reflector might be a better fit for your needs. If you're looking to see a bit of everything and want an all-purpose telescope, a catadioptric telescope—whose focal ratios are generally between those of reflectors and refractors—might be the best bet. They offer sharp-contrast views, are compact, and are suitable for astrophotography too.

The biggest bang for your buck may come from Dobsonian reflectors that offer large apertures, excellent-quality optics, and a sturdy alt-azimuth mount. Sometimes called a "rocker box," the alt-azimuth mount swings the scope up and down for altitude but can also swivel parallel to the horizon. At 6 to 8 inches (150 to 200 mm), in term of aperture, these bargain scopes offer outstanding value for beginners with the lowest cost per inch of any type of instrument on the market. By purchasing them from quality manufacturers like Orion, SkyWatcher, Celestron, and Meade (to name a few), your telescope will also retain good resale value when it comes time to graduate to more serious instruments.

A larger aperture, as offered by this 10-inch (250 mm) lens, brings deep-sky objects into view with improved detail.

USING TELESCOPES

PUTTING YOUR NEW TELESCOPE to use can be rewarding, if you've considered your needs before purchasing. If a smaller, lighter telescope makes it easier for you to examine the night sky on a whim, choose one of those instead of something bigger and more unwieldy to set up and use. A 4-inch (100 mm) telescope will typically allow you to see objects down to magnitude 12, which is more than enough to see all of the planets in our solar system and any number of nebulae, clusters, and galaxies.

Eyepieces

Eyepieces are high-quality magnifying glasses used to examine the image formed by the telescope's main optics. Changing the eyepiece changes the telescope's magnification. Some telescope deals offer a slew of eyepieces yielding high magnifications, but don't be taken in: The reality is that few nights have steady enough skies to let you use more than about 35x telescope aperture. (On a 6-inch/150 mm scope, this would be 210x.) Most of your observing will be done with just two or three eyepieces—probably ones that yield around 50x, 150x, and occasionally 300x.

If the eyepieces that come with your telescope are low quality, you may need to upgrade. Quality manufacturers include Meade, Explore Scientific, and Nagler.

Computerized mounts locate objects and track their movement.

Learn from fellow stargazers at astronomy meetups.

Consult a salesperson to make sure your eyepiece is appropriate, for both the type and size of your telescope. Telescopes can be fitted with a wide range of filters that screw onto the eyepiece, from polarized ones to help resolve details on bright objects like the moon to colored ones that bring out details on planetary surfaces. Resolution is largely dependent on aperture but a filter will augment sharpness or contrast by minimizing the effect of light scattered in the atmosphere or reducing nearby light pollution.

Computerized Mounts

Many new telescopes are dubbed "GoTo" since they come with computerized/robotic mounts that work with planetarium apps on handheld devices. These help sky-watchers keep their favorite sights in view as they move across the sky (or, more accurately, as the earth rotates away from them). They can also help stargazers locate interesting sights in the sky simply by choosing the name of the object they want to look at or just tapping it on a virtual sky chart, which then sends the telescope automatically toward the sky object. Some are compatible with apps that will run off a smartphone.

Some sky-watching purists feel software takes some of the challenge out of hunting down targets, but most find it useful because it quickly locates the best objects the sky has to offer.

Clubs

Local astronomical societies get together on a regular basis to share their enthusiasm for astronomy and the night sky. The membership is a great source of information on telescopes, technical troubleshooting, and everything that's going on in the heavens. You will learn a lot—and will very likely get more out of your telescope and the time you spend using it.

Veil Nebula photographed by the Hubble Space Telescope

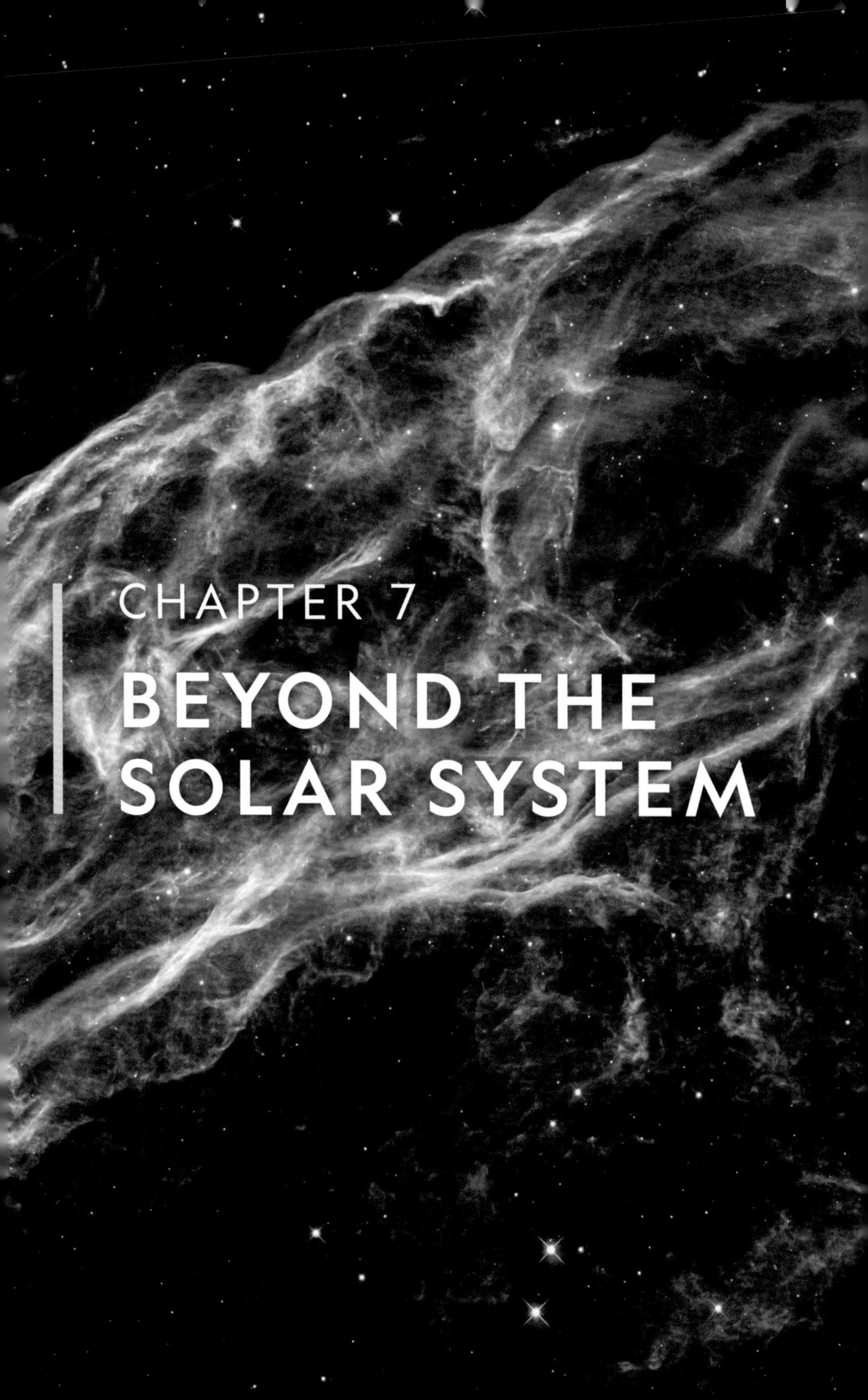

CHAPTER 7

BEYOND THE SOLAR SYSTEM

OUR GALAXY

Long exposures at dark-sky sites reveal depth and detail in the Milky Way.

GALAXIES—THERE ARE TENS OF BILLIONS of them, at least—are among the universe's basic structures: large islands of stars, dust, gas, and mysterious dark matter, held together by their own gravity and congregated in clusters around the cosmos. Our galaxy, the Milky Way, is thought to be about 10 billion years old, a little younger than its oldest stars, which are around 13 billion years old. Different dating methods have produced different estimated ages, but the Milky Way is apparently among the older galaxies in a universe that began 13.82 billion years ago. Its spiral shape is common among larger, brighter galaxies.

Overall, the Milky Way has a diameter of about 100,000 light-years. Its spiral is thickest in the middle—measuring about 13,000 light-years—while its outer arms thin to about 1,000 light-years. Our solar system is located roughly 25,000 light-years from the galaxy core in what is known as the Orion Arm, about halfway to the outer edge.

Beyond the observed objects that make up our galaxy, dark matter—black holes, brown dwarf stars, or an array of exotic objects—seems to be present, if you compare the

FURTHER

The entire spiral structure of the Milky Way spins, completing a rotation about every 226 million years. Our solar system orbits the galaxy's core, traveling at an estimated speed of 143 miles per second (230 km/sec). So in Earth's trip around the galaxy, the last time we occupied the same spot in this galactic orbit was during a time when dinosaurs ruled.

galaxy's visible matter with its speed of rotation. Typically there is not enough visible material to create the gravitational force to hold the system together, meaning that dark matter must be present. Indeed, it may supply the bulk of the galaxy's total mass.

Observation

Cruising the Milky Way with binoculars or a telescope can be spellbinding: Dense star clusters, glowing clouds of gas and dust, decorate this cosmic pathway. Dark nights and clear skies give the best viewing experience of the Milky Way, which can be washed out by even minor sources of light pollution. Earth's ecliptic is tilted nearly 90 degrees to the galaxy, so the view changes seasonally. In the Northern Hemisphere, the summertime perspective is toward the core of the galaxy; the densest region of stars and the brightest in view will be to the south, coursing through Sagittarius. Winter orients the Northern Hemisphere toward the galaxy's outer edge, in Orion and Gemini—a fainter, less star-rich picture. In spring and autumn, we peer beyond the top or bottom of the disk into intergalactic space—a less dense, darker field in which to spot other galaxies.

THE STORY OF the Milky Way

Electricity and urban living have made a Milky Way sighting the exception for much of modern civilization, but it is a testament to the galaxy's dramatic presence in the sky that many cultures developed stories about it. To the Cherokee of North America, it was a trail of cornmeal spilled from the mouth of a godlike dog that was coming to steal grain, until villagers scared it off with their drums. The Seminole saw it as a path leading to heaven—similar to a Norse tale that viewed the Milky Way as the road to their afterlife in Valhalla.

The Milky Way is a spiral galaxy like this one, NGC 4414.

STUDYING STARLIGHT

CONCEPTUALLY, STARS ARE NOT that complicated. Stars are chemically simple, made almost entirely of hydrogen and helium gas. Though the area between stars and galaxies seems empty, the "interstellar medium" is filled with hydrogen, helium, and dust. It is not spread out evenly but occurs in patches, some of which become dense enough for gravity to begin pulling the material inward. If pressure and heat build adequately at the center of the ball of gas, it triggers nuclear fusion: Protons of hydrogen are fused together into helium, releasing an equivalent of around 0.7 percent of the initial mass as energy.

Bayer's Brightness Designations

There are several conventions for naming stars, applied particularly to those included in the major constellations. One of the earliest was created by German astron-

Hot stars in the Lagoon Nebula emit ultraviolet rays.

THE SCIENCE OF Spectroscopy

Visible light falls across a familiar spectrum—the colors of the rainbow, from the longer red wavelengths to the shorter shades of blue. Radio waves, ultraviolet rays, and other types of radiation form part of the much broader electromagnetic spectrum. Astronomers studying distant objects analyze both the visible and invisible portions of the spectrum—a powerful technique known as spectroscopy—by breaking up light into its component colors using a prism. The spectral lines reveal the chemical fingerprints, or composition of a star: its size and mass, how it produces light, and even how fast it's moving. Spectral information can also reveal what material surrounds a star, like the gases in the donut-shaped accretion disks around black holes.

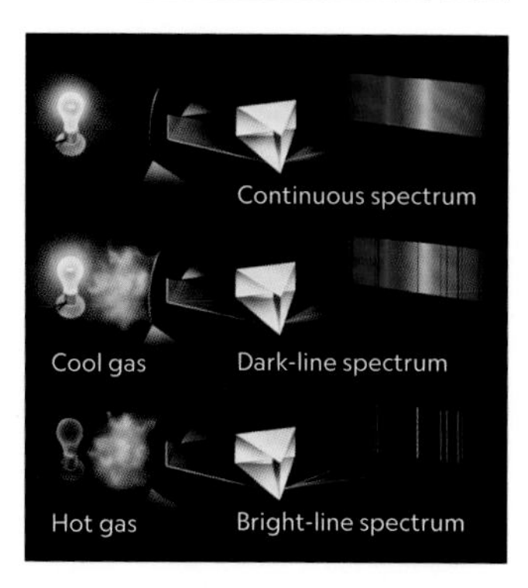

omer Johann Bayer. In 1603 he published the *Uranometria,* which used the 48 traditional constellations as the basis for his naming system. Combining the Greek alphabet with the constellation names, Bayer used the apparent brightness of stars to create a list, with the brightest one in each constellation becoming the "alpha" star, the second brightest the "beta," and on through the 24-letter alphabet (and starting over with capital letters, if necessary). Thus the brightest star in Orion is Alpha Orionis (Latin for "the alpha star of Orion"). Bayer estimated by eye, so improved technology has shown that his labels aren't perfect—sometimes an alpha star isn't the brightest. In the case of Orion, Beta Orionis ranks higher in magnitude than the alpha star. However, the Bayer designations are commonly found on star guides and charts, including in Chapter 10 of this book, and it's a helpful system for defining and discussing the objects under study. Many stars were named by Arab, Greek, or Roman astronomers and carry "proper names" such as Betelgeuse (Alpha Orionis) or Sirius (Alpha Canis Majoris), colloquially called the Dog Star. There is a lot more to stars than their brightness, so several other catalogs of names exist to index the millions of stars identified by astronomers.

FURTHER

The first star catalogs were put together by Chinese astronomers in the fourth century B.C. Today contemporary lists have cataloged millions of stars brought into view by modern equipment. The Hubble Space Telescope Guide Star Catalog, for example, includes around 19 million stars.

THE STELLAR FAMILY

BRIGHTEST STARS IN THE NORTHERN HEMISPHERE

SIRIUS
CONSTELLATION: Canis Major
SEASON: Winter
MAGNITUDE: −1.44
DISTANCE (LY): 8.6

ARCTURUS
CONSTELLATION: Boötes
SEASON: Summer
MAGNITUDE: −0.05
DISTANCE (LY): 36.7

VEGA
CONSTELLATION: Lyra
SEASON: Summer
MAGNITUDE: 0.03
DISTANCE (LY): 25.0

CAPELLA
CONSTELLATION: Auriga
SEASON: Winter
MAGNITUDE: 0.08
DISTANCE (LY): 42.9

RIGEL
CONSTELLATION: Orion
SEASON: Winter
MAGNITUDE: 0.18
DISTANCE (LY): 864.3

PROCYON
CONSTELLATION: Canis Minor
SEASON: Winter
MAGNITUDE: 0.40
DISTANCE (LY): 11.5

BETELGEUSE
CONSTELLATION: Orion
SEASON: Winter
MAGNITUDE: 0.45
DISTANCE (LY): 642.5

TWO SIMPLE TRAITS—color and brightness—were used to create a star classification system a century and a half ago. Today we have hundreds of thousands of stars that have been sorted and cataloged using the same methods.

Star Varieties

The science of spectroscopy was a breakthrough in understanding the variety of stars in the universe. In the late 1800s, a Harvard University team began classifying stars based on a spectral analysis of their hydrogen emission lines. One astronomer who joined the team, Annie Jump Cannon, combed through tens of thousands of photographs of star spectra to devise a simple classification scheme. The star listing was named the Henry Draper Catalog, after the astronomer who began the project. By the end of Cannon's work it included about 225,000 stars, including some 50 times dimmer than the faintest stars visible to the naked eye.

The scale began as an alphabetical sequence that matched spectral class with color. But when Cannon arranged spectra by the strength of their hydrogen lines (which reflected temperature), she found the sequence ran from hot, blue O stars to cool, red M ones, with classifications B, A, F, G, and K in between. More than a century

An artist's interpretation of a blue supergiant

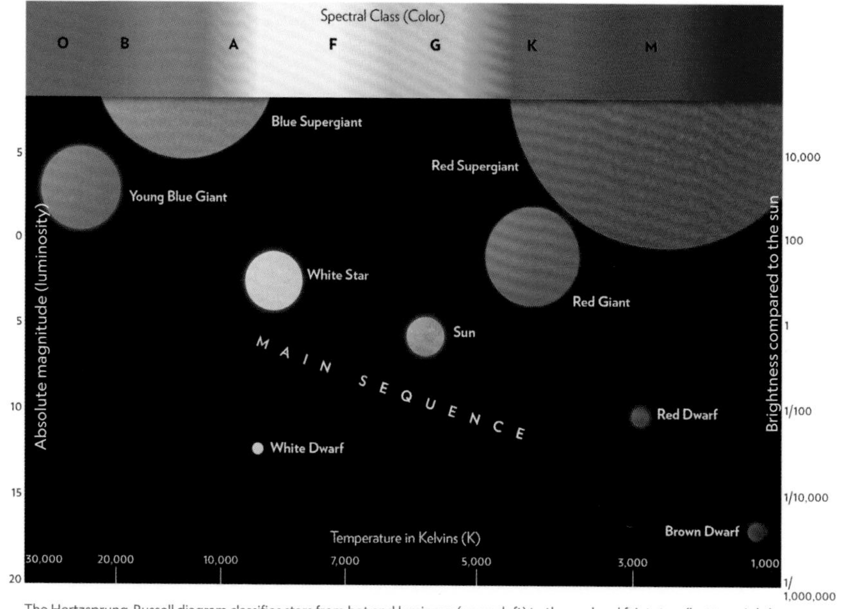

The Hertzsprung-Russell diagram classifies stars from hot and luminous (upper left) to the cool and faint stars (bottom right)

old, the scale has proved an efficient way to convey with a single letter a star's temperature and chemical composition. The cool M-type stars burn at around 5000°F (2760°C), for example, and their spectra indicate the presence of metals like magnesium and titanium oxide. O stars burn at more than 10 times that temperature and show the presence of ionized helium, carbon, and oxygen. Each category is subdivided further on a 0 to 9 scale based on temperature, from hot to cold.

Hertzsprung-Russell

Even as Cannon continued her work, Ejnar Hertzsprung of Denmark and Henry Norris Russell of the United States began independently plotting stars' color and temperature against their luminosity to reveal their properties. The main sequence running diagonally through the Hertzsprung-Russell diagram reveals order in the relationship: The hotter, bluer stars are also larger and brighter, while the cooler, red dwarf stars are smaller and dimmer.

FURTHER

Brown dwarfs have the ingredients to become stars, but they lack adequate mass for nuclear fusion. Only a few times larger than Jupiter, brown dwarfs are dubbed by astronomers as "failed stars" for how they give off significant heat but don't shine.

STELLAR ZOO

FURTHER

The easiest binary stars to observe are those that have a wide enough separation for your binocular or telescope to "split" the stars. Remember that the limit of your ability to separate binaries into individual stars is determined by your instrument's aperture size and the atmospheric conditions. The larger the main objective lens/mirror, and the drier and steadier the air mass above your observing site, the better your resolution capabilities. A small telescope, even at low magnification, will reveal the stunning Albireo star system, where one star shines blue and the other gold.

AS WE LEARN MORE about stars and their structure, it seems our solo-traveling sun may be an outlier. The massive gas clouds that nurture stars usually produce them in pairs or multiples that remain gravitationally attached to each other throughout their lives. Binary stars, triples, and larger groups are the norm, whether the companion is a twin of similar size and luminosity or a group of small siblings that stick close to the dominant member of the system.

Stars Orbiting Stars

It was not until the late 1700s that astronomers began distinguishing so-called optical doubles—coincidental alignments of stars that only appear to be close—from true binary star systems, with two stars orbiting a common center of gravity. In the latter, a telescope is almost always needed to pick out the individual members. Indeed, the partners in one of the most famous pairs of stars in the sky—Mizar and Alcor—are about one light-year apart yet are linked by gravity.

In 1650 Italian astronomer Giovanni Battista Riccioli—famous for his maps of the moon and his opposition to heliocentrism—observed that Mizar had a close companion star of its own that was unrelated to Alcor. He'd made the first discovery of a binary star. Since then, astronomers have found that both components in the Mizar system are also binaries, making Mizar a quadruple star system. By the late 1770s, enough double stars had been located that British astronomer William Herschel began a deliberate hunt for them. Lists of known binary and multiple stars, including the observations of amateur binary hunters like Sherburne Wesley Burnham, were eventually compiled into lists like the 17,180-entry Aitken Double Star Catalog.

Albireo double star in Cygnus is a good target for telescopes.

Seeing Binary Stars

Powerful binoculars will bring some

THE SCIENCE OF Variable Stars

Variable stars change luminosity, sometimes by a large enough degree to change whether or not they can be seen by the naked eye. That is the case with Mira in Cetus, which disappears and reappears on an 11-month cycle. Mira-type stars are red giants that have reached an unstable point in their lives and then start to pulsate. Rarer Cepheid variables pulsate on a regular cycle as they expand and contract over a period of days or weeks. Another type of variable star, called T Tauri stars, fluctuate at random rather than on a fixed period. T Tauri-type variables are still being formed, and change in luminosity as they contract and shed outer layers of gases. Some of the more prominent variable stars to observe include Eta Aquilae and Mira (Omicron Ceti).

A sequence of images shows how variable star V1 in the Andromeda galaxy glowed brighter over a few weeks.

of the more prominent binaries—Epsilon Lyrae and Beta Cygni, for example—into view, but you will need a telescope to tackle more subtle double stars. Binaries will appear close together in the field of view. Distance, differences in magnitude, and other factors, however, can affect your ability to distinguish the "B" star from its brighter and more dominant "A" companion. Gamma Andromedae will produce an orange-blue pair; a close look at Theta Orionis in the Orion Nebula will yield, depending on the power of the telescope, four to six stars in a configuration known as the Trapezium.

LIFE CYCLE OF STARS

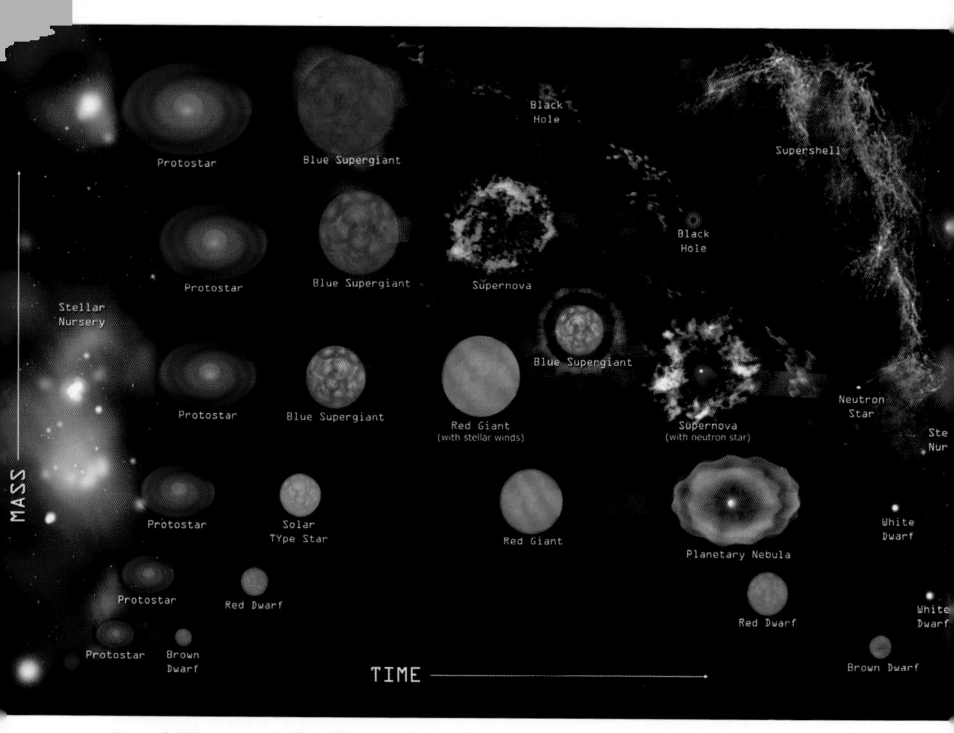

Stars of different mass follow different cycles of birth, middle age, and eventual demise.

THE MASSIVE AMOUNT OF GAS contained in a star is, nevertheless, finite: The nuclear furnace at its core will eventually run out of fuel, and the star will die. How long its life is and how dramatically it ends depend on the star's mass.

Because of the energy needed to sustain them, large stars consume their hydrogen fuel supply at a relatively faster rate than smaller ones. The sun, considered a medium-size star, formed with enough fuel to live for about 11 billion years. Far more massive blue giants might burn out in a quick million years or so, while smaller red dwarfs—the most common type of star—will saunter along for tens of billions of years. The smallest of these dwarves may continue to burn unchanged perhaps for trillions of years before they

FURTHER

The deaths of old stars in supernova explosions help create new stars as they throw newly forged, heavier elements back into space.

burn out, a period that extends longer than even the lifetime of the universe.

For most of their lives, stars exist along the main sequence of the Hertzsprung-Russell diagram, acting in a stable and predictable way. Hydrogen atoms fuse into helium, energy is released, and the outflow of energy offsets the inward tug of gravity—the system is in balance. But as stars begin to run out of fuel to burn, they start to defy that basic connection and may jump out of the main sequence. A white star, for example, might balloon into a red giant, becoming cooler and redder yet brighter as it swells enormously in size. Stars like the sun will eventually collapse into a hot white dwarf.

How a Star Dies

As the fuel tank starts to run dry, a star's nuclear core begins to shut down. The crush of gravity can no longer be held off, and the star collapses in on itself. Temperatures rise, igniting hydrogen in the outer layers of the star and initiating a new round of fusion in which helium atoms (the by-product of hydrogen fusion) are forged into carbon. At this point, gravity is overcome by a new surge of energy and the star begins to expand.

Average-size stars grow into pulsing red giants (like the bright stars Arcturus in Boötes and Aldebaran in Taurus). Once the available fuel is completely consumed, they then shed their outermost layers of atmosphere, exposing the remains of the hot core as they transition to what is called a white dwarf—a spent, stable, and gradually cooling ember. In the process, the cast-off gas forms a cloud around the former star that we call a planetary nebula. These expanding gas shells glow with fluorescent light and can last for tens of thousands of years before they fade. Famous examples include the Dumbbell, Helix, Cat's Eye, and Ring nebulae.

Supergiant stars with a mass four to eight times as massive end their life by exploding into a violent supernova. The outer layers are shed to leave only the small dense core that gravity will condense into a high-density neutron star. The most massive supergiants have such a large gravitational source that they collapse all the way to a black hole.

FURTHER

Large gas clouds—literal star nurseries—have been found scattered throughout the Milky Way. You will find a prominent one by looking through binoculars or a telescope at the middle of the three stars that form the "sword" that hangs from the belt of the constellation Orion (see page 264). A small telescope will reveal several young stars in the Orion Nebula, and studies of the area indicate that others are now forming. The raw material for star formation can also come from the supernova explosions of older stars across the universe.

EXPLOSIONS & BLACK HOLES

THE END STAGES of a high-mass star's evolution can result in a series of spectacular events. Novae and supernovae are gargantuan explosions—the stellar equivalent of atomic bombs. When supergiant stars explode, they sometimes collapse into black holes, which are among the universe's truly strange and awesome events.

FURTHER

The universe's supply of gold and other heavy elements is forged during supernova explosions and during the bizarre, but spectacular, merger of two neutron stars. In 2017 one of these stellar collisions was detected 150 million light-years away, and it created a huge amount of gold in the host galaxy, estimated to be up to 13 times Earth's mass.

Novae

Typically, a nova occurs when stars in a binary system reach different points in their evolution—with one collapsed into a dense white dwarf and the other in a red giant phase. If the red giant expands far enough, the white dwarf's gravity may begin siphoning hydrogen from its companion. As the cloud of hydrogen accumulates around the dead star and becomes more massive, pressure and temperature build until the stolen gas erupts in a flash of thermonuclear fusion. The event may brighten the star by as many as 10 magnitudes—sometimes bringing it into naked eye or binocular view for weeks or months—until the remnants of the blast dissipate. This whole cycle can continually repeat every few decades or centuries.

Seeing a Supernova

Supernovae, by contrast, are onetime events. Type II supernovae occur in binary stars if the white dwarf draws

The Crab Nebula is the remnant of a Type II supernova.

A black hole leaks radiation and particles.

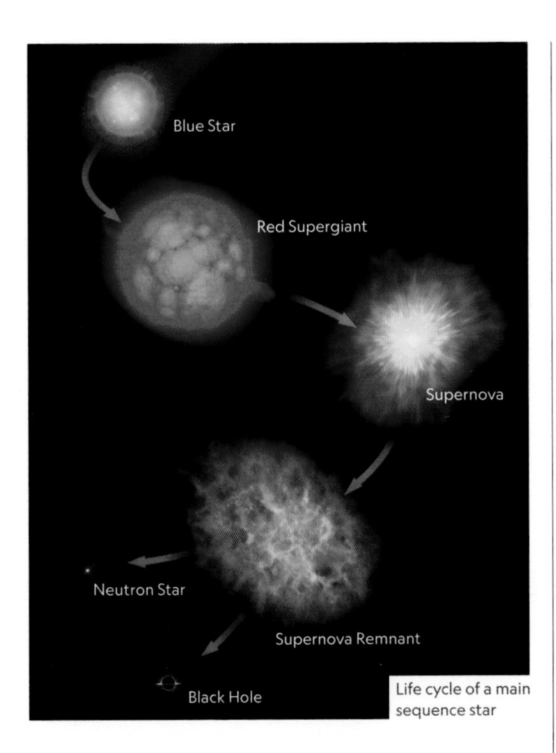

Life cycle of a main sequence star

in so much hydrogen that its core implodes. Type I involves an implosion at the core of a red supergiant star, which unleashes a violent shock wave that blasts away the surrounding cloud of gas. Supernovae unleash so much energy that the star involved briefly shines as bright as its entire home galaxy. Supernovae can occasionally be seen by the naked eye—events that have astonished earthbound observers. Backyard telescope users regularly make supernovae discoveries in neighboring galaxies.

Black Holes

Although some supergiant stars collapse into neutron stars at the end of their lives, the largest supergiants produce such massive gravitational forces that matter literally folds in on itself and collapses into extremely dense matter called a black hole. Nothing can escape its gravitational pull—not even light.

SKY-WATCHERS
Stephen Hawking

Renowned British astrophysicist Stephen Hawking, who battled motor neuron disease for more than five decades, spent his life studying the universe's strangest and most awesome events surrounding black holes. His theory that black holes are not really black, and in fact leak radiation and particles, is now an accepted concept in physics—one that solved the puzzle of how these gravitational pits could exist and still be theoretically joined with our conventional understanding of energy. The discovery of their heat emission is known as Hawking Radiation. Hawking showed that black holes can act as test beds for attempting to unite the theories of large and small physics: squaring Einstein's theory of gravity with the quantum theories of electromagnetic and nuclear forces, which has long been a battleground in the world of physics.

STAR CLUSTERS

FURTHER

The Pleiades star cluster, known as the Seven Sisters, contains about 3,000 stars. Most beginners can easily catch sight of the five brightest stars unaided—Alcyone, Atlas, Electra, Maia, and Merope. Some records suggest that up to 20 Pleiades can be spotted under very dark skies using averted vision: Look around the object and not directly at it.

DEEP-SKY TERRITORY is the domain of open star clusters and globular clusters. These dramatic formations are sometimes visible to the naked eye, but a telescope will give viewers a deeper sense of the universe's scale and complexity. There is no perspective quite as breathtaking.

Open Clusters

The dusty spiral arms of the Milky Way galaxy are filled with and defined by thousands of groupings of stars, many of them gravitationally bound into what are known as open star clusters. Some appear to harbor a few dozen stars, while others are impressive collections of thousands, filled with stars that shared the same birth clouds and huddle together across a few light-years of space. Groupings like the Pleiades, in the constellation Taurus, have been recognized for centuries. Formed as stars coalesce

Open clusters like the Beehive cluster (M44) contain hundreds of stars.

out of giant molecular clouds of gas and dust, these clusters move together through the Milky Way and are held together—temporarily, at least—by their mutual gravitational pull. They are usually around the same age but vary in terms of mass, contained in well-defined areas of space measuring between 10 and 20 light-years across.

Globular Clusters

Globular clusters also share a common origin, but they are several orders of magnitude larger. These massive balls of stars—as many as a million—are typically 100 light-years across. Globulars are considered old-timers, believed to have formed 10 to 12 billion years ago from the same ancient material that helped form the Milky Way when the universe was only a fraction of its current age. Unlike open clusters, which contain mostly young blue stars, most globulars are populated by red giants.

About 150 globular clusters have been identified, but dozens may still be hidden from sight on the other side of the galaxy. They appear to congregate mostly in the outer reaches of the Milky Way, strewn about in a halo above and below the plane of the galaxy, centered around the galaxy's core like bees buzzing around a hive. They are best spotted in regions of the sky that offer clear views away from the rich star clouds of the Milky Way, in constellations like Hercules, Sagittarius, and Ophiuchus. It's possible to observe them with the naked eye, but a 4-inch (100 mm) telescope will begin to resolve the outer stars and a 10-inch (250 mm) instrument will show a grainy profusion down to the core. They can be very densely packed, which makes identifying individual members at their centers difficult. This compactness is an illusion, however, as these core stars are actually separated by a light-year or more.

CLUSTERS

PLEIADES OPEN CLUSTER
CONSTELLATION: Taurus
SEASON: Winter
MAGNITUDE: 1.20
DISTANCE (LY): 442

BEEHIVE OPEN CLUSTER
CONSTELLATION: Cancer
SEASON: Spring
MAGNITUDE: 3.10
DISTANCE (LY): 577

DOUBLE CLUSTER
CONSTELLATION: Perseus
SEASON: Autumn
MAGNITUDE: 3.7/3.8
DISTANCE (LY): 7,502

M13 GLOBULAR CLUSTER
CONSTELLATION: Hercules
SEASON: Summer
MAGNITUDE: 5.78
DISTANCE (LY): 22,180

WILD DUCK OPEN CLUSTER
CONSTELLATION: Scutum
SEASON: Summer
MAGNITUDE: 5.80
DISTANCE (LY): 6,197

Globular clusters like the Hercules cluster (M13) contain hundreds of thousands of stars.

FURTHER

In 1974, a radio signal was beamed from the Arecibo radio telescope to globular star cluster Messier 13, some 24,000 light-years away. It was one of the first and most powerful radio broadcasts ever sent, addressed to any extra-terrestrial beings who might be listening.

INTERSTELLAR CLOUDS

Emission nebulae glow with the light of newborn stars.

A NEBULA IS A STELLAR CLOUD that is part of the interstellar medium, or the gas between stars. And while the dramatic images in astronomy books are, for backyard sky-watchers, mostly restricted to images created with a telescope, live views through the eyepiece still give a sense of the universe's dynamics. You can spot about two dozen nebulae with the naked eye: faint, glowing patches suspended in a sea of stars. Ancient astronomers took notice of these faint fuzzies and called them *nebulae,* Latin for "clouds." While many of these interstellar clouds are invisible to the human eye since there is nothing to light them up, many are detected either by the radiation they emit or by the light that passes through them. Their visibility depends a lot on their origin, how they're lit, and what elements they hold.

Emission Nebulae

Emission, or glowing, nebulae are clouds that glow with the light of newborn stars nestled within, which heat their gases to extremely high temperatures around 10,000 degrees Kelvin. These stars emit intense ultraviolet radiation that bathes the entire nebula, causing its gases to re-emit this energy as light and glow like a neon sign such as the nebula that appears to be the middle star of Orion's sword.

FURTHER

Some of the largest, densest nebula-type objects within galaxies are what are called giant molecular clouds—stellar nurseries composed of vast reservoirs of hydrogen gas. Each one can hold enough gas and dust to give rise to thousands of stars.

Despite their name, planetary nebulae have nothing to do with planets. They're actually emission nebulae left behind when a dying red giant sheds its outer layers, such as the Ring Nebula in Lyra. Many such nebulae are visible in a 4- to 6-inch (100 to 150 mm) telescope. Use higher magnification for better contrast and experiment with light pollution filters to enhance visibility.

Dark and Reflection Nebulae

Dark nebulae are clouds filled with gas and dust so dense it blocks any visible or ultraviolet light from passing through. As a result they appear as dark, irregularly shaped silhouettes set against a rich background of stars. Despite being invisible to the human eye, some appear bright in infrared, signifying an internal heat source that's likely from internal collapse and star formation. The result can create spectacular visual effects such as the famous Horsehead Nebula.

As their name suggests, reflection nebulae do not shine under their own power but reflect the light from nearby stars. The dust within these cosmic clouds scatters the nearby star's light, making it glow. And because the tiny dust grains scatter light at the blue end of the spectrum more easily, astronomers find reflection nebulae tend to take on eerie shades of blue.

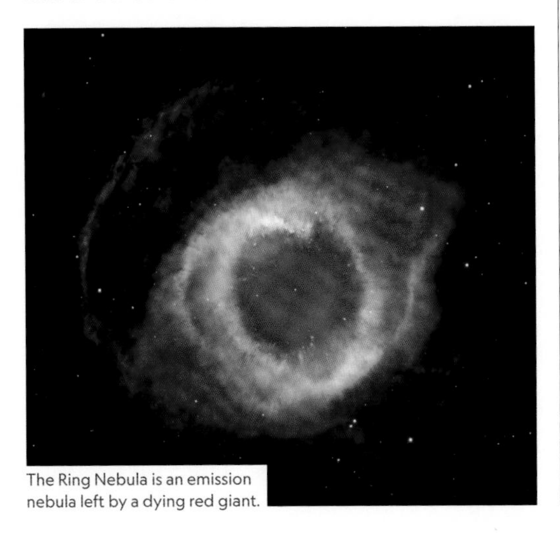

The Ring Nebula is an emission nebula left by a dying red giant.

THE SCIENCE OF Hubble Color Pictures

The Hubble Space Telescope is a true astronomical discovery machine and a window out onto the universe, producing iconic imagery that has captivated our imagination while offering a feast for our eyes. But are the dramatic colors we see real? Turns out the orbiting observatory doesn't see in color, but instead snaps raw grayscale images. It takes several images of the same object using multiple filters, each of which lets different light wavelengths through, that are then combined into a single composite. Since many nebulae emit light that is too faint for the human eye to see, colors are sometimes enhanced to highlight subtle structural features that would be invisible otherwise.

DEEP-SKY OBSERVING

ABD AL-RAHMAN AL-SUFI, the 10th-century Persian astronomer, showed that even the naked eye can carry us beyond the Milky Way. His records identify "the little cloud" in the constellation Andromeda—what we now know to be the Andromeda galaxy, 2.6 million light-years away. Granted, Al-Sufi was looking at pristine skies free of the influence of electric light. Clear, dark skies are crucial for modern-day sky-watchers: Even under ideal conditions, distant galaxies just barely outshine the background glow of the night.

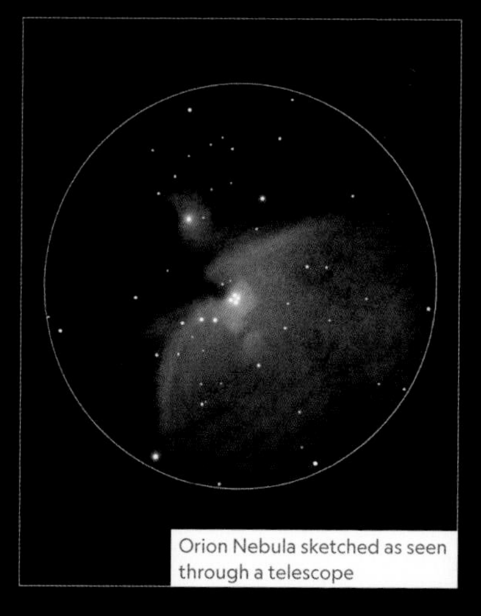

Orion Nebula sketched as seen through a telescope

Viewing Tips

Light pollution is one of the most critical factors when it comes to observing clusters, nebulae, and galaxies—even more than telescope size. Within city limits, artificial lights wash out the subtle details in these dim, diffuse objects that often take up a relatively large part of the sky. Visually, the Lagoon Nebula in Sagittarius and the Beehive cluster in Cancer both take up the space of three times the full moon's disk. We can't see the full extent of this magnificent gas-and-dust cloud and group of stars because the light is stretched across such a large swath of sky, making a darker sky necessary to get the full picture.

Orion Nebula glowing above Mt. Hood

Being aware of sky conditions also makes a difference in your viewing experience. Moonlight produces its own annoying light pollution for deep-sky observers, so plan your deep-sky hunts on or around the new moon each month to ensure dark skies.

You can get sharper views of objects when they are positioned higher up in the sky, near their zenith. That's because there is a thinner column of air for the light to travel through and get washed out. It's best to time your observing sessions in terms of optimum time of night, or season, when the object reaches its highest point in your local sky. The deep-sky objects in Chapter 10 are pinpointed on

Orion constellation with blue Rigel (lower right), orange Betelgeuse (upper center), and nebula in between

the seasonal sky charts to indicate what time offers the best viewing.

It's worth remembering that smaller telescopes, when used under pristine country skies, can better showcase deep-sky objects than a larger scope in a brightly lit suburban backyard. But generally speaking, when it comes to clusters, the larger the telescope, the more details you will be able to resolve—especially individual stars.

Hints of Color

One of the most common disappointments for beginners viewing deep-sky objects is the lack of obvious features and the absence of the brilliant colors seen in photographs. Those colors are added later to reveal light from wavelengths that are out of range of human vision. Telescopes can't give us those same vivid details and colors, but there are tricks you can use to bring out hidden details. Look to the side of an object instead of directly at it, a practice known as averted vision, and give your scope a very slight shake: That will make structural details stand out more clearly.

The human eye is just not sensitive enough to pick up colors from deep-sky objects, except for the few that have a high surface brightness. Along with a few bright, compact planetary nebulae, the Great Orion Nebula is one some people can view color. While it looks like a hazy patch of light to the naked eye, an 8-inch (200 mm) or larger telescope will reveal an eye-catching swirl of clouds with hints of green and pink.

Spiral galaxy M74 dotted with pink regions of hot hydrogen gas

CHAPTER 8

BEYOND THE MILKY WAY

COSMIC DAWN

THE STORIES ancient cultures used to explain the cosmos sound strange today, but are they any stranger than the reality? Consider this: An infinitely small point of infinite density and gravity erupts to create not just space but also time, unleashing a torrent of radiation that cools and starts to form matter, which starts to form stars, galaxies, planets—and us. Based on estimates from the Hubble Space Telescope, that universe contains perhaps 120 billion galaxies, each containing billions of stars.

Fundamental questions about our universe remain unanswered (and perhaps unanswerable), but recent discoveries have created a more detailed picture of how it has evolved.

FURTHER

A team of Australian researchers has estimated that the total number of stars exceeds 70 sextillion (that's 70 times a billion times a trillion—a seven with 22 trailing zeros). The team, which used two of the world's most powerful telescopes, surveyed one strip of sky.

Age of the Universe

The existence of this radiation marked a turning point in the debate over the universe's origins. Trying to determine the source of static corrupting satellite signals, Bell Labs engineers Arno A. Penzias and Robert W. Wilson, in 1965, famously determined that the "noise" was coming from the entire sky, at a wavelength consistent with background radiation from the big bang. It is this most ancient remnant of light that offers a glimpse into the earliest first moments of the universe.

A false color image of a galaxy cluster shows starlight in orange, hot gas in green, and the core of dark matter in blue.

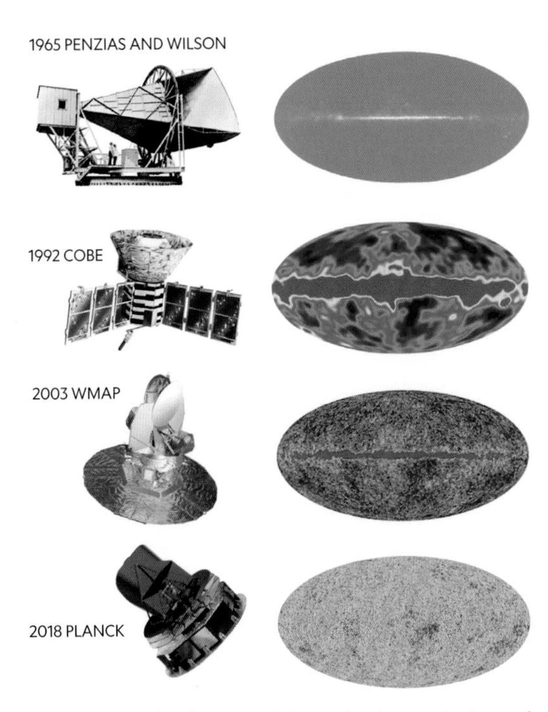

Improved technology has mapped the cosmic microwave background of the universe in increasing detail.

The clock can't yet be pushed back to "time zero," but recent data from the Wilkinson Microwave Anisotropy Probe (WMAP) has led to what scientists regard as an accurate estimate of the universe's age. The WMAP, launched in 2001 into an orbit more than 1,000,000 miles (1,600,000 km) from Earth, developed a detailed map of cosmic background radiation, the echo of the origin of the universe. They measured differences in the temperature of that background radiation across the sky. The variance, though minuscule, explains how heavier elements, stars, and galaxies arose out of an early cloud of radiation and matter. The WMAP temperature map is consistent with the emergence, over 13.77 billion years, of a universe that looks like ours. In 2013 the European Space Agency's Planck spacecraft refined the measurement to 13.82 billion years.

THE SCIENCE OF Cosmic Radiation

It is a general property of materials that they tend to heat up when they are compressed and cool when they expand. Thus, we would expect that in its earlier stages, the universe was hotter than it is now, simply because it was smaller and more compressed. In other words, we would expect that the universe had a hot beginning and has been cooling ever since. We have found evidence for this ancient heat in the Cosmic Microwave Background (CMB), radiation left over from the big bang. Invisible to the human eye, this cosmic background radiation is detectable today mostly as very faint microwaves. Measuring tiny differences in temperature amid this radiation is how we know roughly how old the universe is, and that much of it is composed of dark energy and dark matter—even if we don't yet understand what they are.

LOOKING BACK IN TIME

FURTHER

The Hubble Space Telescope has provided information about galaxies and stars as they existed near the beginning of time. In a universe calculated to be 13.8 billion years old, it collected light from a superbright galaxy 13.4 billion light-years away.

LIGHT HAS A FINITE SPEED—186,000 miles (299,000 km) per second. The implications of this are easy to grasp on a local level. Since the sun is about 93 million miles (150 million km) from Earth, the light leaving its surface takes about 500 seconds to reach our eyes. In other words, we see the sun as it was 8.3 minutes ago; moonlight shining through your bedroom window left the moon 1.3 seconds ago. But carry that over a galactic scale and the results are mind-boggling. Light spends so long traveling to Earth that scanning the skies is equivalent to looking back in time. Train a backyard telescope on the Andromeda galaxy—at 2.5 million light-years away our nearest galactic neighbor—and you are literally peering back in time, seeing that collection of stars at it was about the time *Homo habilis* emerged in Africa.

The Expanding Universe

The universe is expanding, changing the positions of deep space objects. How do we know? Because as space objects

Dwarf galaxy the Large Magellanic Cloud is around 170,000 light-years away.

A primordial quasar as it may have appeared in the very early universe

change their position relative to Earth, their color appears to shift. Much like with sound, light works on a spectrum, with waves that can get stretched out or scrunched up, changing frequency from low to high, depending on our relative position to them. Think of how an ambulance siren seems to get higher in pitch as it approaches you, then gets lower and deeper as it moves away—something called the Doppler Effect. In space when an object moves away from us, its wavelength gets longer and its frequency lower, which we can observe because its light shifts into the red part of the spectrum (called redshift). Those getting closer to us have a shorter wavelength, moving into the blue part of the spectrum (called blueshift).

Redshift is a key piece of evidence for the big bang. It's also important in helping astronomers figure out and compare the distances between faraway objects and understand the structure of the universe. For sky-watchers, some of the most challenging observing lies in locating these distant, redshifted galaxies, but just being aware of their jaw-dropping distance makes them rewarding to hunt for.

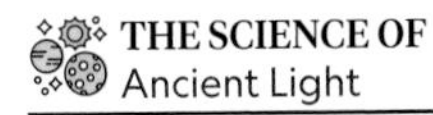

THE SCIENCE OF Ancient Light

Because light takes time to travel through space, the farther away an object is from us, the longer it takes for its light to reach us. That means we only ever see it as it appeared sometime in the past. In the case of the most distant galaxies, we see them as they were billions of years ago. But could there be galaxies so far away that their light hasn't reached us? Because we appear to live inside a universe that is continually expanding at an accelerating rate, there may be far-flung galaxies we can't see—and never will. To see them, we would have to break Einstein's special theory of relativity and travel faster than the speed of light, which would require an infinite amount of energy.

THE BIG BANG

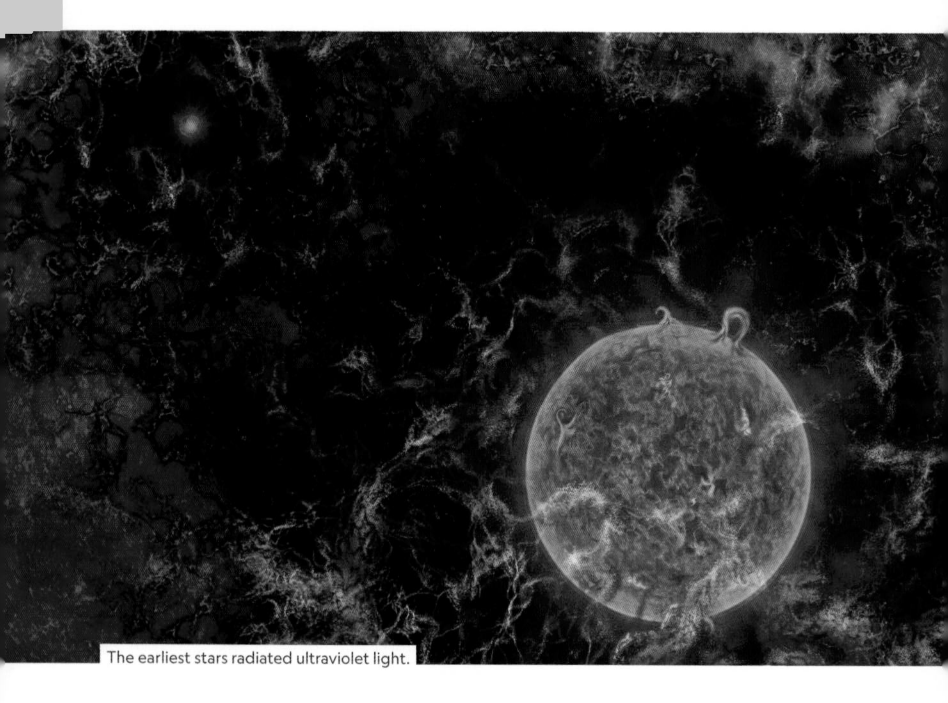

The earliest stars radiated ultraviolet light.

SIR FRED HOYLE'S INVENTION of the term "big bang" was certainly unintentional. The English physicist championed a "steady state" cosmology, in which matter was constantly being generated to fill an expanding universe—a stable process that could go on indefinitely. He coined the term out of sarcasm to mock those who felt the universe began in a single cataclysm, but the name stuck. Hoyle's insult now refers to a theory that, while hardly perfected, has stayed consistent with what we've detected as we peer ever deeper into space.

Universal Origins

If current thinking is correct, the universe began as a dense, hot point of matter where space, time, and all subsequent matter and energy were compressed into an object the size of a pea. The concept has led theoretical physicists to try to explain how such entities could exist, yielding to exotic constructions like string theory—the idea that everything is made of massless filaments of energy. Those filaments would exist in more dimensions than the four (three in space and one in time) current theories can account for.

Current cosmology has not yet pushed back to that moment of origin, but it does try to explain the chain of events that unfolded less than a second afterward. In the first trillion-trillionth of a second after the big bang, theories suggest that the universe expanded with incomprehensible speed. It has continued to do so, but much more slowly, over the following billions of years. Initially (in this case, that's a span of about a million years) these elements existed within a fog of radiation.

In this beginning, there were gamma rays—high-temperature, high-energy photons. As the cataclysm unfolded and space-time expanded, the gamma rays cooled and decayed into basic particles of matter. At about the 300,000-year mark, however, free atomic particles were increasingly absorbed into stable atoms. The simplest atoms—hydrogen and helium—formed in the roughly three-to-one ratio found in the most ancient observed stars. This matter continued to condense into the first stars and galaxies we see today.

When matter condensed, the universe was no longer cluttered by free-wheeling electrons, so physicists say that the universe became "transparent." It could be permeated by light as well as some microwaves, which has allowed devices like satellite TV as well as the WMAP and Planck to measure the cosmic background radiation: the leftover imprint glow detected throughout the universe from the original light emanating from the big bang event.

Matter vs. Antimatter

During the early stages of the big bang, both matter and antimatter particles formed—opposites that, when they collided, annihilated each other in a huge burst of energy. Antimatter and matter particles should have formed in equal numbers, but because of a slight surplus matter won out, leaving behind our visible universe. How that antimatter was destroyed, or if it exists elsewhere in the universe, is one of the unsolved issues for big bang theorists.

THE SCIENCE OF the First Stars

Until the first stars began to shine, the universe was a dark and featureless expanse. These "dark ages" lasted a mere 180 million years after the big bang. In 2018, astronomers in Australia used a tabletop radio telescope to investigate when the very earliest stars may have turned on. The infant universe that led to these primordial stars was devoid of any light, filled with neutral hydrogen gas that was impossible to detect. Over time, clumps of gas condensed into ever-larger, compact clouds that ignited at their cores and formed the first stars. Their ignition released intense ultraviolet radiation into the dark void of space for the first time, and this radiation interacted with the surrounding hydrogen gas. Astronomers suggest that the gas then absorbed this onslaught of stellar fallout and began to glow with radio signals. It is this faint radio emission from hydrogen gas that this small radio instrument was able to detect, and its very existence indicates the birth of the cosmos' first stars.

NIGHT SKY ENIGMAS

THE AGE OF THE UNIVERSE is only one of its riddles. As we push farther into space, mysteries have presented themselves in the form of things that challenge our understanding of what's out there, including dark matter, dark energy, and quasars.

Dark Matter

Observations of supernovae in the 1990s indicated that the universe is expanding at an accelerating rate—faster than earlier theories predicted, leading to the premise that "dark energy" is forcing galaxies away from each other, countering the effects of gravity. Roughly 70 percent of the universe is made up of mysterious dark energy, whose existence can only be inferred.

While dark energy pushes the universe apart, dark matter is a powerful gravitational force that draws objects together. In the current phase of discovery, their relationship is not yet clear. Matter as we see it around us is thought to make up only about 4 percent of the universe. Scientists have estimated that about 26 percent of the universe consists of cold dark matter: objects we can't see, and whose presence is witnessed only because light from distant galaxies bends in a way that reflects the gravitational pull of an unseen mass. By comparing the amount of estimated matter in a galaxy with the gravitational force

FURTHER

Even exotic objects like quasars are within reach of the skilled amateur with the right equipment and a keen eye. In the constellation Virgo (see page 202) just a few degrees north of the star Porrima, or Gamma Virginis, quasar 3C 273 can be spotted through a backyard telescope with at least an 8-inch (200 mm) aperture.

The Hubble telescope reveals the most distant galaxies ever observed.

Light is emitted from a quasar, the brightest object discovered in the universe.

needed to hold it together, scientists have concluded there's something missing, as the mass of visible matter is not enough to keep it from ripping apart. So-called dark matter supplies that missing mass: Brown dwarf stars, black holes, and other massive compact halo objects (MACHOs) play a role, as may the still hypothetical entities known as weakly interacting massive particles (WIMPs).

Quasars

Quasars were first detected in the 1960s from their radio emissions. At first, there was no visible object attached to these emissions. When the source was found, these distant objects appeared to be a normal star, yielding the term "quasar" for quasi-stellar radio source.

By calculating the redshift of the light coming from quasars, we discovered they couldn't be producing it with ordinary stellar fusion—they were too far away for light to reach Earth at the magnitude they were producing it. So theorists turned to black holes, whose interactions with surrounding stars and gases create light emissions perhaps a trillion times as bright as the sun—so, the only thing powerful enough to keep quasars humming along. The Sloan Digital Sky Survey Quasar Catalog has identified hundreds of thousands of quasars so far but they are many millions or billions of light-years away.

THE SCIENCE OF Quasars

Quasars are monstrous, ancient baby galaxies that look to us just as they did back in the early universe. Each one is powered by a supermassive black hole that weighs billions of times the mass of our sun at its heart. As the black hole consumes the contents of nearby stars, brilliant streams of energy, dust, and gas are released from above and below it in the form of jets that accelerate to near the speed of light. They are the brightest objects in the universe, shining up to 100,000 times brighter than our Milky Way. The Hubble telescope discovered that these intense energy sources are part of normal galactic development. They were violently active during the first few billion years of the universe, but over time the flow of energy decreased and they evolved into the "normal" galaxies seen today.

GALAXIES

Elliptical galaxy

Spiral galaxy

Barred spiral galaxy

Irregular galaxy

OUR MILKY WAY and other galaxies are made of stars, interstellar dust, gas, and dark matter (not fully understood, but that includes objects such as brown dwarf stars, planets, black holes, and exotic objects that do not emit light). Galaxies come in three basic shapes—elliptical, spiral, and irregular.

Galactic Shapes

Elliptical galaxies, which make up about 18 percent of all galaxies, may be nearly spherical or be stretched into an oblong. Elliptical galaxies are composed mostly of older, red giant stars that follow their own paths around the galactic center. Spiral galaxies rotate around a bright nucleus, with arms spiraling out from that point. Seventy-eight percent of all observed galaxies are spirals, which hold a mix of old and young stars and typically have dusty regions of new star formation. At least one-third of all spiral galaxies are known as barred-spirals, like the Milky Way, that have spiral arms sweeping out from the galactic core.

Irregular galaxies, which represent about 4 percent of

FURTHER

The Milky Way and Andromeda galaxies may have formed closer together before a collision with a dwarf galaxy drove them apart.

the total, are amorphous collections of stars that include star-forming nebulae but lack any coherent shape.

Collisions

Galaxies are so massive that the strength of one can rip apart another, and even in the relative emptiness of space they can collide. For example, in the constellation Corvus, the Antennae galaxies are in the process of merging. In some cases, larger galaxies may simply absorb nearby dwarf galaxies. Apparently, our own Milky Way galaxy has cannibalized many smaller satellite galaxies.

Clusters

Containing billions of stars, galaxies are themselves arrayed in clusters around the universe. The Local Group, for example, includes the Milky Way and Andromeda as the dominant members, and more than 30 much smaller neighbors scattered over a distance of about 10 million light-years. Other clusters have thousands of members, and the clusters themselves form into larger structures.

Superclusters also seem to have a shape, grouping into large-scale structures that give the overall universe a "clumpy" nature. Galaxies, clusters, and superclusters seem to be arrayed in strands much longer than they are thick.

THE SCIENCE OF the Great Attractor

As if a black hole at the center of the Milky Way was not worrisome enough, the deep-space "Great Attractor" may have its own surprises. Located some 150 million light-years away, this aggregation of galaxies and dark matter is drawing in everything around it, including our Local Group of galaxies. We are being pulled toward it at roughly 500 miles (805 km) a second, as is the rest of the Virgo supercluster. A cluster of galaxies known as Abell 3627 has been pinpointed by space-based observation as the center of a massive supercluster, which is causing the "attraction."

The grand-scale structure of the universe is a web with clusters and superclusters of galaxies along filaments and knots.

GALAXY-HUNTING GUIDE

GOING DEEP INTO THE SKY requires practice and a willingness to use all the tools at hand. The magnified views offered by binoculars and telescopes are critical in bringing structural details into focus. Keep in mind the trade-off between magnification and field of view in your choice of a telescope eyepiece: Spotting large galaxies requires lower power, a wider field of view, and a reliance on contrasts; smaller, brighter objects can be seen under greater magnification, provided you can capture it in the narrower field.

Observing through a Dobsonian telescope

Navigating the Deep Sky

Observing galaxies through backyard telescopes can be challenging for both beginners and seasoned sky-watchers. Galaxies are large, extended objects spread out over a significant portion of the sky—perhaps many degrees, or many moon disks, wide. Spiral arms, which have a low surface brightness, are barely visible even in large-aperture telescopes. To get the best views, you will need dark, transparent skies—especially when it comes to finding face-on spirals, which tend to appear as hazy oval patches. Spirals that have their disks tilted to our line of sight are easier to spot.

Larger, 12- to 16-inch (300 to 400 mm) telescopes will begin to tease out structural details: dark dust lanes, or perhaps the galaxy's outer regions, which sometimes show some mottling. These subtle, irregular patches indicate star-forming regions, home to nebulae and star clusters.

The tried-and-true way to hunt down galaxies is by jumping from star to star, from field to field of view on a clear path. The sequence takes us from wide- to narrow-angle views, becoming more and more magnified and filled with ever-fainter stars. Start with the naked eye, then move to binoculars or finderscopes, then from wide-angle, low-power eyepieces to higher power ones. To see nebulae and galaxies, use a finderscope that has 8x to 10x magnification with a 50 mm-wide lens. They are perfect for faint star hopping and also for spotting galaxies with low surface brightness.

Treasure Troves

The Messier catalog is considered home to 40 galaxies, which are some of the biggest and brightest the sky has to offer a backyard observer. Depending on sky darkness, three can be seen with the naked eye: M31 (the Andromeda galaxy), M33 (the Triangulum galaxy), and M81 (Bode's galaxy). But all of the ones on the Messier list can be spotted with nothing more than 7 × 50 magnification binoculars under good sky conditions, which makes it perfect for the novice galaxy hunter.

The next set of galaxies worth exploring are part of the New General Catalog (NGC) and its Index Catalog (IC). As many as a few hundred galaxies are within reach of an 8-inch (200 mm) telescope, while a 16-inch (400 mm) scope will bring that number into the thousands.

In the constellation Leo, 6-inch (150 mm) telescopes will offer many easy-to-locate galaxies, while 8- to 12-inch (200 to 300 mm) telescopes will keep you busy in the bowl of the Big Dipper, the Great Square of Pegasus, and Hercules's Keystone.

Generally, telescopes up to 6 inches (150 mm) can see galaxies with surface brightness of magnitude 13, while a 10-inch (250 mm) will go down to magnitude 14. Just keep in mind that the galactic core is usually the brightest and easiest to observe.

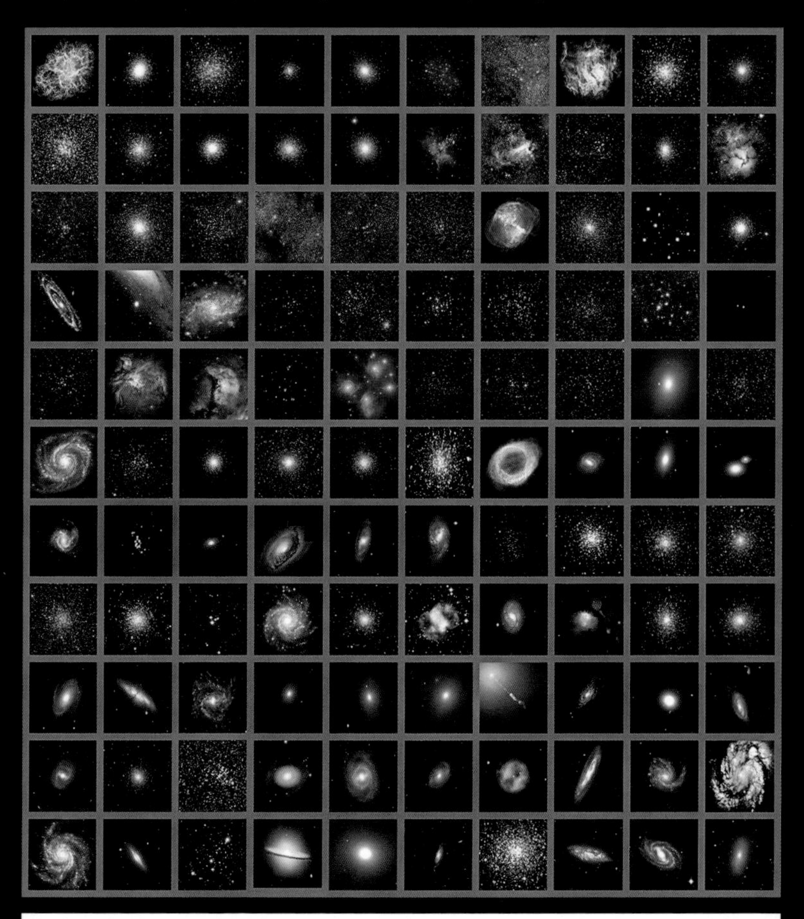

The deep-sky objects cataloged by astronomer Charles Messier start with Crab Nebula M1 (top left) and end with the dwarf elliptical galaxy M1110 (bottom right).

Comet Hale-Bopp streaks through the sky in 1997.

CHAPTER 9

NAVIGATING THE NIGHT SKY

CONSTELLATIONS & ASTERISMS

Leo in the stars and in mythological artwork

IN THE CAVES OF LASCAUX, France, prehistoric frescoes drawn some 16,000 years ago include what researchers believe are elements of an early sky map. There are what appear to be diagrams of the phases of the moon, references to the Pleiades star cluster, and an illustration of the three-star group we know today as the Summer Triangle. These drawings are evidence of a human urge to order the heavens—whether for purposes of tracking the seasons, guiding travel and navigation, or conveying history through stories. The process has been handed down through the patterns known as constellations and asterisms. It's important to know the difference between the two.

Constellation Origins

Forty-eight constellations were identified in antiquity, including large and well-known star patterns such as Orion, Scorpius, and Pegasus. Incorporated into poems like Aratus's "Phaenomena" in the third century B.C. and codified in Ptolemy's *Almagest* in the second century A.D., the names in use today reflect their origins in Greco-Roman mythology. Yet by giving a larger identity to individual stars, particularly the brighter ones, constellations have helped create a map of the heavens that is still used today.

FURTHER

The constellation Leo is one of the oldest known star patterns, recognized by many cultures as the figure of a lion. Ancient Mesopotamians recorded it as early as 4000 B.C.

It was not until the beginning of the 1500s that increased travel from the Northern to the Southern Hemisphere brought areas of the sky into view that were unknown to the Babylonians, Greeks, and Romans, and with them came a whole new set of constellations added to their catalogs.

Ultimately, the list grew to include 88 star patterns, including a handful of new northern ones, filling in the whole of the south celestial polar region.

During the 1920s, with the pace of discovery in astronomy quickening, the International Astronomical Union (IAU) formalized the boundaries of each constellation so that each represented not just a group of stars but a roughly rectangular patch of sky. That constellation-driven map is now a basic tool for astronomers to orient themselves in the night sky. People may draw in the connecting lines in different patterns but the member stars and the section of sky are consistent.

Asterisms

Don't be surprised when you look at a list of constellations and find some prominent names, like the Big Dipper and Little Dipper, missing. These are asterisms: small groups of stars that have a distinct, well-known shape and may form part of a constellation but are not considered constellations in and of themselves. The Dippers, for example, are part of Ursa Major and Ursa Minor.

The Summer Triangle composed of brilliant blue Vega (top left), Deneb (lower left), and Altair (right), with the Milky Way cutting through the middle

THE STORY OF
The Southern Sky

Polynesian voyagers populated the southern skies with constellations whose positions relative to each other and to the horizon were used for wayfinding. Meanwhile, pre-Columbian nations in ancient Mesoamerica made advanced studies of the sky as evidenced by their calendars and the astronomical alignments in their monumental architecture.

NORTHERN & SOUTHERN SKIES

FURTHER

The traditional list of constellations began to expand in 1595, when Dutch navigator Pieter Dirkszoon Keyser, on an expedition to the East Indies, suggested a dozen new star formations, all added to Johann Bayer's 1603 atlas of the heavens, the *Uranometria*. The trip would have taken him around the tip of Africa, at roughly 34.5 degrees south latitude.

WHEN PTOLEMY CODIFIED the constellations, he was using traditions most likely handed down from Mesopotamian observers who had watched the skies at about 35 degrees north latitude. The northern sky up to the celestial north pole would have been visible. Theoretically, they could have looked over a 90-degree span to the south as well—or down to a declination of around minus 55, though the practical limit would be perhaps 10 degrees less than that, since objects near the horizon are difficult to see. Anything farther south would have been blocked by Earth. Some southern features such as Crux, the famous Southern Cross, crept over the horizon but not enough to make a full impression: Until the 17th century it was included as part of Centaurus.

Your Viewpoint

Just like the ancients, your view of the constellations will be determined by your latitude. Though constellations are often characterized as belonging to the Northern or Southern Hemisphere, it helps to envision them on a continuum. Using the celestial coordinate system, each constellation occupies a range of declination—akin to terrestrial latitude—on a hypothetical celestial sphere that encases Earth (see page 15). Those with more northern declinations on the 0 to 90-degree scale will be higher and more visible for more of the year in the Northern Hemisphere (likewise in the south for those with negative, southerly declinations from 0 to −90).

Engraving of the southern constellations from 1708

Indeed, at the extreme, some constellations (Ursa Minor, Cepheus, and Draco, to name a few)

Crux, the Southern Cross, is a prominent constellation in the southern sky. The dark patch to its left is the Coalsack Nebula.

are "north circumpolar"—they are at such a high northern declination that they are visible in the northern sky throughout the year, never dipping below the horizon. Some of these may peek into the south during certain times of year, but just barely, visible only to people within a few degrees of the Equator. The Big Dipper appears to just barely peak above the northern horizon from northern parts of Australia. Some prominent constellations, such as Orion and Virgo, are positioned so close to the celestial equator that they can be seen from both hemispheres on a seasonal basis. In between the circumpolar constellations and the equatorial ones, what you see depends on where you are. Viewers at the Equator can theoretically see all 88 constellations over the course of a couple of nights of viewing. But it's worth noting that the orientation of familiar constellations can appear tilted, sideways, or even upside down when viewed at vastly different latitudes. For instance, Orion will appear tipped on his head as viewed in the Southern Hemisphere by travelers coming from the north.

FURTHER

Home to some of the world's monster astronomical observatories, Chile's Atacama Desert has become one of the hottest destinations for professional and amateur sky-watchers. With daytime trips to various giant telescope facilities and nighttime tours of the unpolluted starry skies, astrotourism has become a booming business.

COSMIC SIGNPOSTS

Bright naked-eye signposts direct to fainter telescope targets.

LEARNING THE CONSTELLATIONS takes a bit of patience, but it requires no equipment other than our eyes. Remember, the major star patterns were first identified by the naked eye thousands of years ago. They can occupy a large portion of the sky, so a wider field of view is essential. While a visit to dark sky country can leave beginners bewildered as they try to pick out unfamiliar reference points from a very crowded backdrop, a moderately light-polluted suburban backyard may be ideal, in a sense, for locating the very brightest reference stars and using them to find landmark constellations.

Constellation Calendar

From a typical urban location, you may find the sky in one or more directions blocked by backyard trees or nearby buildings. Determine which vista is open and locate sky charts that will show you what's in that part of the sky over the course of the year. In the mid-northern latitudes, for example, with a reasonable portion of the southern sky open, you can enjoy summer views of Sagittarius and

Scorpius, then watch as Orion begins to dominate the sky in autumn and winter.

Be sensitive to the cyclical aspect of the constellations. While the few that are circumpolar will theoretically always be visible, they will also, depending on latitude, dip lower in the sky over the course of the year. They, too, have a season when they are highest and most visible. Constellations at their zenith, their highest point in the sky, will be free of atmospheric disturbance and appear most clearly. To plan your year of stargazing, you will need a sky chart. Planetarium apps will allow you to pick a date and time and set coordinates, as will websites like *Stellarium.org*. Monthly magazines will also do the trick, as will a variety of books and other publications.

FURTHER

Many songbirds migrate at night, when there are fewer predators, and use the stars to help them navigate on their journeys. Bird species like buntings use Polaris to help orient themselves, tracking down its location by observing the sky's rotation around it. Studies have shown that other bright stars in constellations such as Orion also help guide the internal compass of some migrating species.

Starry Signposts

Typically, charts will use Bayer designations—a Greek letter and a Latin constellation name—to designate a constellation's component stars in order of magnitude. But not all "alpha" stars are created equal—designations only refers to the most luminous star relative to the other members of that constellation. The reason Orion stands out is that it includes three of the brightest stars in the sky—Rigel, Betelgeuse, and Bellatrix. Canis Major includes Sirius, the very brightest, as a reference point, while Gemini has the bright stars Castor and Pollux. Most likely, these will be indicated on your chart by large circles. Look for the biggest, determine when those constellations will come into your field of view, and spend some time with them. Then you'll be prepared to star hop your way to fainter star patterns.

Even some of the signs of the zodiac require proper orientation. Cancer, for example, has no star brighter than magnitude 4. But it rests between Gemini and Leo, two show-offs with bright stars that can help locate the scuttling crab.

Cassiopeia is a powerful signpost in autumn.

KEY STAR PATTERNS

THE NIGHT SKY IS FULL OF WONDERS, and those wonders change over the course of the year as the constellations and other star patterns slowly make their way across the heavens. Major seasonal stellar signposts include the Summer Triangle asterism (summer) and the constellations Cassiopeia (autumn), Orion (winter), and Leo (spring). To help you get started in your explorations, let's take a quick visit to two of the most well-known and easy-to-find star patterns: One you can see all year long and one to look for in northern winters.

The Big Dipper

The Big Dipper is probably the most familiar star pattern in the northern sky, in part because it's so far north that most people in the Northern Hemisphere see it all year long. But what many people don't know is that it's not actually a constellation. It's an asterism: a pattern of stars that is not one of the 88 constellations officially recognized by the IAU. Asterisms can be composed of stars entirely within an already recognized constellation, such as the Sickle in Leo, or of stars from more than one constellation, as is the case with the Summer Triangle, formed from stars in Cygnus, Aquila, and Lyra. The Big Dipper is contained within an official constellation: Ursa Major (the Great Bear).

Orion

Star Hopping From the Big Dipper

The Big Dipper's handle is located in the Great Bear's tail, while the square body of the Dipper is the bear's hindquarters. Drawing imaginary lines connecting various star pairs within it will guide you to neighboring constellations. Extending an imaginary line north from orange-tinted Dubhe will lead you to Polaris. Following the curve of the Big Dipper's

Orange-tinted Dubhe in the "cup" of the Big Dipper points to Polaris.

handle outward leads to Arcturus in the constellation Boötes (and the brightest star in the northern sky); continuing along the curve leads to Spica, the brightest star in the constellation Virgo. There's even a saying about this among stargazers: "arc to Arcturus and speed on to Spica."

Orion (the Hunter)

In the Northern Hemisphere, Orion is the lord of the winter sky, and his distinctive shape is filled with bright stars and other astronomical sights. Orion features two of the brightest stars in the sky: To the north is Betelgeuse, the 10th brightest star; to the south is Rigel, the seventh brightest. But the real action, astronomically speaking, is in Orion's belt and the "sword" that hangs from it. There you will find the Orion Nebula, one of the few nebulae that is easily seen with the naked eye (although seeing its beauty at its best requires a telescope). Orion is also a wonderful stellar guidepost. Extending imaginary lines out from the belt stars leads to neighboring constellations Canis Major and Taurus.

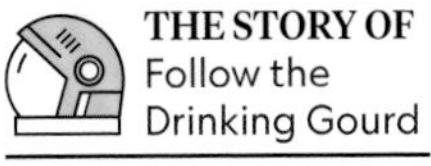

THE STORY OF Follow the Drinking Gourd

The Drinking Gourd is another name for the Big Dipper asterism. According to American folklore, this star pattern aided slaves in their journey to freedom on the Underground Railroad by pointing the way to the North Star.

THE ZODIAC

As constellations were added to Ptolemy's list, it turned into a grab bag of items—a compendium not just of myths and monsters but also of microscopes and sextants and other objects that reflected the scientific sensibility of the 17th and 18th centuries. But the most ancient star patterns remain at the core and form a sort of foundational first step to learning the sky. These are the constellations of the zodiac: 12 well-known animals, people, and creatures. They appear as a belt around Earth along a path tightly bound to the ecliptic, the same projected path that the sun, most planets, and the moon trace annually across the sky.

Star Signs

The names of the zodiacal constellations are familiar to most, tied up as they are in the pseudoscience of astrology and in pop culture. The word "zodiac" comes from the Greek *zodiakos kyklos*, or "circle of animals." Most do have animal associations: Aries (the Ram); Taurus (the Bull); Gemini (the Twins); Cancer (the Crab); Leo (the Lion); Virgo (the Virgin); Libra (the Scales); Scorpius (the Scorpion); Sagittarius (the Archer, a centaur); Capricornus (the Sea Goat); Aquarius (the Water Bearer); and Pisces (the Fish).

Some of the original associations have been lost to history. The twins in Gemini

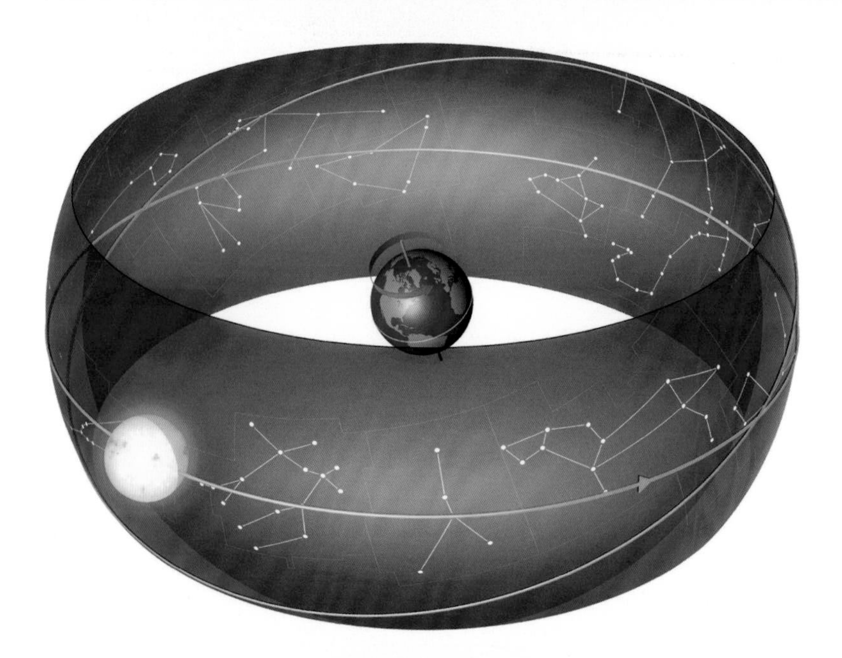

The sun appears to pass through the 12 zodiacal constellations.

are usually linked with Castor and Pollux of Greek mythology but have also been pairs of peacocks and goats. Capricornus may have started as a simple goat—a common sight to Chaldean or Babylonian herdsmen—but "grew" a fish tail, a possible reference to a Greek tale in which the god Pan flees a monster by jumping into the Nile. Other civilizations, like the ancient Chinese, have entirely different figures in their version of the zodiac, many of which are religious characters. The ecliptic may have first been divided into 12 signs by the Babylonians as far back as 3000 B.C.

Capricornus on a 17th-century sky chart

Phases

What truly sets the zodiac apart from other constellations is its position along the ecliptic: the path of Earth's orbit around the sun, hence the illusion of a wheel. Over a year, as the planet moves through its orbit, each of the 12 will follow in succession, coming into view in the east, rising higher night by night, and then lowering into the west until that patch of sky has disappeared for the year. It is this westernmost phase, when the sun is roughly aligned with, or "in," a particular constellation, that is associated with the signs of the zodiac. Sagittarius, for example, after spending the northern summer months prominent in the night sky, has by late fall shifted to the west. The sun is roughly in front of it, and by twilight in the north it has almost disappeared below the horizon. And so it "controls" the period from November 22 to December 21.

Astrological Mismatch

While the fabled zodiac does describe 12 constellations, there are actually many more—up to 21, perhaps—straddling the ecliptic. Ophiuchus, the Serpent Bearer, is the largest of these forgotten constellations. Owing to its enormous size in the sky, the sun actually appears to spend more time gliding across Ophiuchus than it does better-known Scorpius; however, for unknown reasons, it was left out of the astrological zodiac. Adding to the confusion is the fact that Earth's axis wobbles like a gyroscope, which means the sun's path through the constellations has changed since the ancient invention of the zodiac. While the sun may have been within a certain constellation on a certain date of the year in ancient times, it is now in a totally different part of the sky on that same date.

A clocktower unveiled in Venice, Italy in 1499, depicts the Ptolemaic system of the universe with the sun orbiting Earth and passing through the zodiacal constellations.

CHAPTER 10
SKY CHARTS

SEASONAL SKY CHARTS

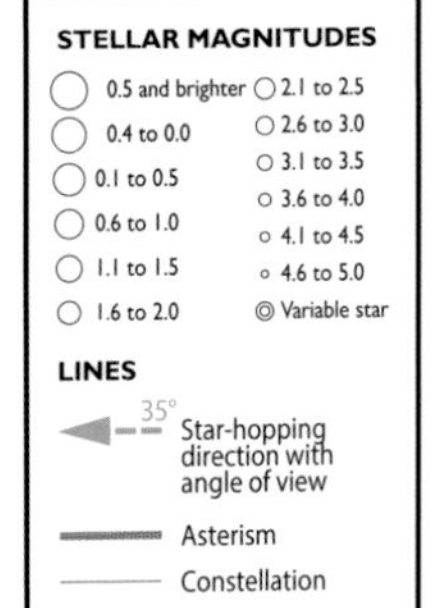

THE PAGES THAT FOLLOW provide a basic guide to the sky in the Northern Hemisphere—your road map to the stars. The maps are organized by season: Each has its own set of two circular all-sky maps as well as focused sky charts and descriptions depicting individual constellations best seen in that season.

The first chart for each season maps the visible constellations, brightest stars, and locations of more accessible deep-sky objects, including star clusters, nebulae, and galaxies. The second seasonal chart maps the same sky, but it is a simplified view that focuses on easy-to-find asterisms (or star patterns) around the brightest constellations, with arrows to help you hop from one part of the sky to another. Star hopping is a key technique stargazers use to navigate from the brightest objects in the sky to the fainter ones (see page 20).

Remember, the celestial sphere operates on a constantly changing continuum: The positions shown on these charts

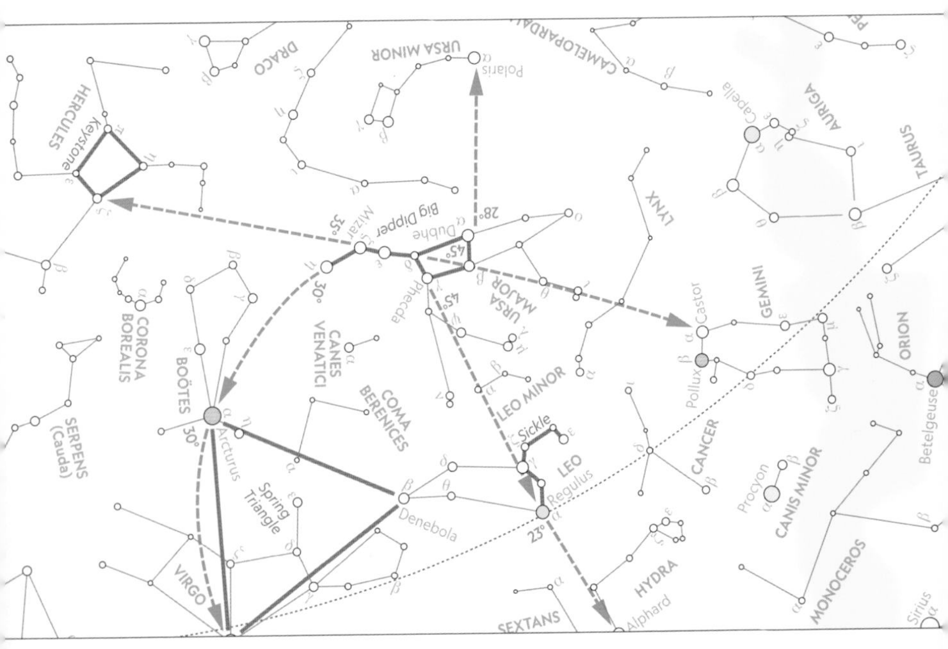

Sample star-hopping chart

apply to particular days, times, and latitudes. These maps are from the perspective of someone at 40 degrees north latitude, with the dates and times indicated on each chart.

Getting Oriented

These maps are approximations of what you'll see at a particular time, but they will still prove useful on any night. Like the sun, the stars appear to move from east to west, in some cases falling from view below the horizon for several months. Turn the map around so that the compass direction at the bottom of the circular seasonal chart is the same direction as you are facing. Observers at a latitude above 40 degrees north will find the northern constellations higher in the sky, for longer periods of time, while losing sight of some constellations below the southern horizon. The reverse is true as a position moves toward the Equator. To illustrate, take a look at the maps for the North Star, Polaris, which does not rise, set, or change position during the year. It is roughly in the middle of the northern sky—reflecting the orientation of 40 degrees north latitude. Observers farther north will see it higher in the sky. By contrast, it will be lower in the sky as your point of observation moves farther south. Other stars will shift accordingly.

Map Features

The maps include a silhouette line around the border indicating the horizon. Although constellations near the horizon are visible hypothetically, anything within about 10 degrees of that edge will be difficult to spot.

Stars of magnitude 5—easily visible from countryside locations—have been connected by lines to form the constellations. Stars of magnitude 3.5 or brighter are labeled, and some have been tinted to reflect their color. A star's size reflects its magnitude, though keep in mind that fewer stars will be visible with the naked eye from light-polluted cities than those shown in these pages. Deep-sky objects are indicated with different symbols (displayed in the accompanying legend). The ecliptic—Earth's path around the sun, and the line of travel for the zodiacal constellations—is represented as a dotted line across each map; the faint, white field represents the Milky Way.

BEST VIEWING TIMES

AURIGA
STAR CHART: Winter
MONTHS: Dec./Jan.

ORION
STAR CHART: Winter
MONTHS: Jan./Feb.

TAURUS
STAR CHART: Winter
MONTHS: Dec./Jan.

CANCER
STAR CHART: Spring
MONTHS: Mar./Apr.

BOÖTES
STAR CHART: Summer
MONTHS: May/July

SAGITTARIUS
STAR CHART: Summer
MONTHS: July/Aug.

LYRA
STAR CHART: Summer
MONTHS: July/Sept.

CAPRICORNUS
STAR CHART: Summer
MONTHS: Aug./Sept.

CYGNUS
STAR CHART: Summer
MONTHS: July/Sept.

PEGASUS
STAR CHART: Autumn
MONTHS: Sept./Nov.

CONSTELLATION CHARTS

ONCE YOU ARE FAMILIAR with the sky at large, you are ready for a more detailed look at individual constellations. Here we have them organized into seasonal sections, depending on when they are best viewed. Within those sections, they appear in order of their prominence in the night sky and the interesting objects contained in them.

This chapter includes 58 of the 88 recognized constellations, including the original 48 described in Ptolemy's *Almagest* and some newer constellations that were added later to "fill in" the sky. Those not included are the "deep south" constellations that cannot be viewed from mid-northern latitudes, presuming that the best season to view each constellation is the time of year when it appears highest in the sky during the evening hours for

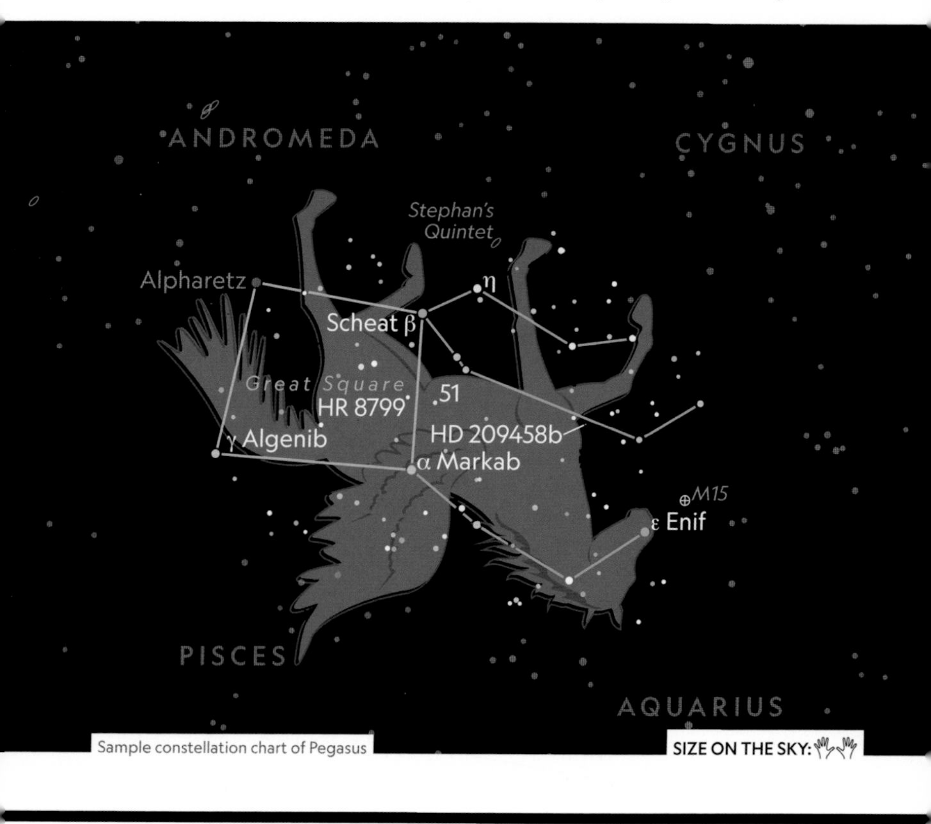

Sample constellation chart of Pegasus

observers at 40 degrees north latitude, the parallel that runs near New York City; Columbus, Ohio; Denver, Colorado; Salt Lake City, Utah; and Northern California.

Fast Facts

Underneath each constellation's name, you'll find a short list of key items and facts. "Makeup" refers to the number of brighter stars (down to magnitude 5) found in that particular constellation—all visible to the naked eye under dark skies and in good viewing conditions. "Best Viewed" indicates the best months to hunt for it. "Location" indicates where the constellation appears in the seasonal all-sky chart, the direction in which to look to find it in the night sky. The "Deep-Sky Objects" are galaxies, double stars, nebula, or other sights that make good targets for binoculars or a telescope. Where appropriate, enhanced telescope images show interesting objects in the constellation and sidebars describe them in greater detail.

Mapped Constellations

Every constellation features a star map showing the main constellation, the interesting objects within it, and neighboring night sky objects. "Size on the Sky" uses outstretched hands and closed fists to indicate the constellation's approximate size in the sky (see page 20). The backdrop shows background stars and nearby constellations to help orient your view. Lines connect the constellation's member stars, and a shaded illustration shows an interpretation of the mythological character for which it is named. Prominent asterisms are also named.

Frequent "Main Star" sidebars provide facts about the most important stars in a constellation, detailing their mass, color, and distance from Earth. They are ordered by their apparent brightness which usually, though not always, aligns with the Bayer system, with the brightest star generally given the Greek letter alpha, the next brightest beta, and so on. Well-known stars are labeled with their proper names and, where appropriate, tinted to indicate their approximate color. For fainter constellations, only the brightest star is labeled.

SPRING

SPRING IS GALAXY-HUNTING SEASON. Earth's position relative to the Milky Way shifts so that the core of the galaxy lies just below the horizon in the east. We look out into deep space with a better view of elusive sky objects. As the nights become shorter, Orion sets in the west and is nearly lost in the twilight glow just as Boötes, its bright orange star Arcturus, and Hercules climb higher in the east.

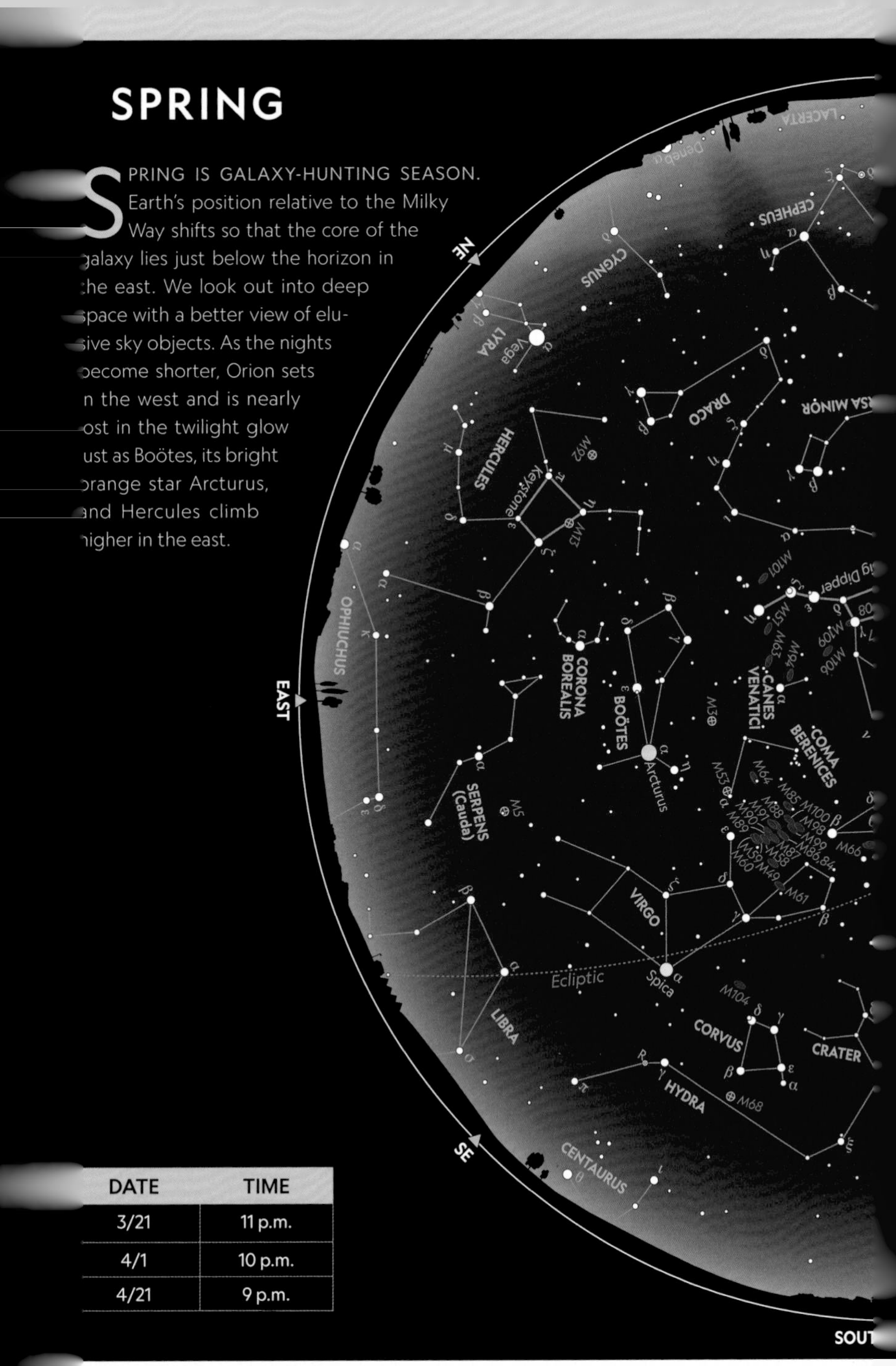

DATE	TIME
3/21	11 p.m.
4/1	10 p.m.
4/21	9 p.m.

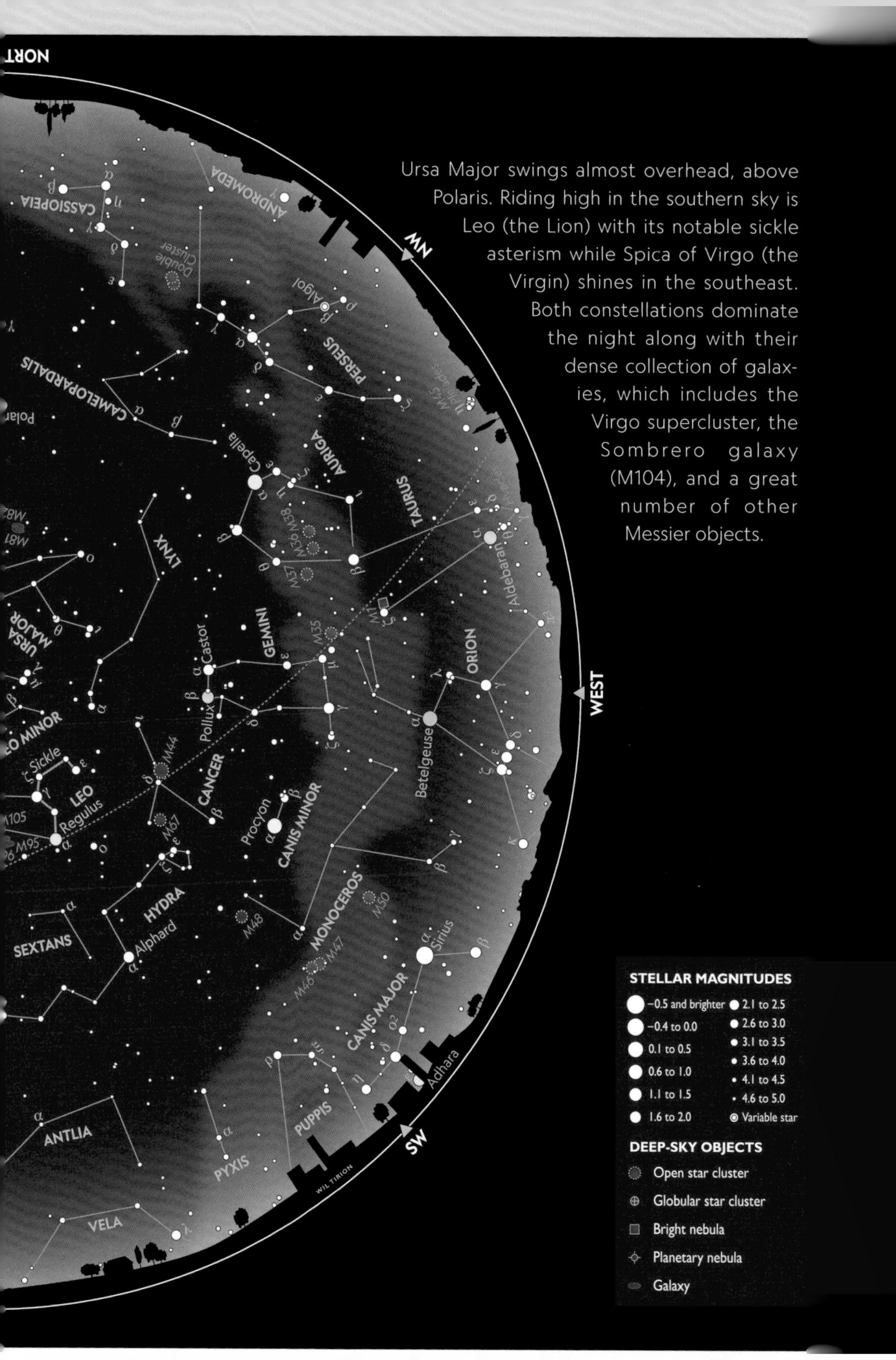

Ursa Major swings almost overhead, above Polaris. Riding high in the southern sky is Leo (the Lion) with its notable sickle asterism while Spica of Virgo (the Virgin) shines in the southeast. Both constellations dominate the night along with their dense collection of galaxies, which includes the Virgo supercluster, the Sombrero galaxy (M104), and a great number of other Messier objects.

SPRING STAR HOPPING

AS THE GREAT BEAR prowls the sky overhead, the Big Dipper asterism points the way, in nearly every direction, to nearby constellations. Drawing an imaginary line down south from its bowl leads to blue-white star Regulus in Leo. Continuing down toward the horizon, this line goes on to point to Alphard of Hydra. By using other combinations of Big Dipper stars, observers can find Hercules and Gemini throughout the spring season.

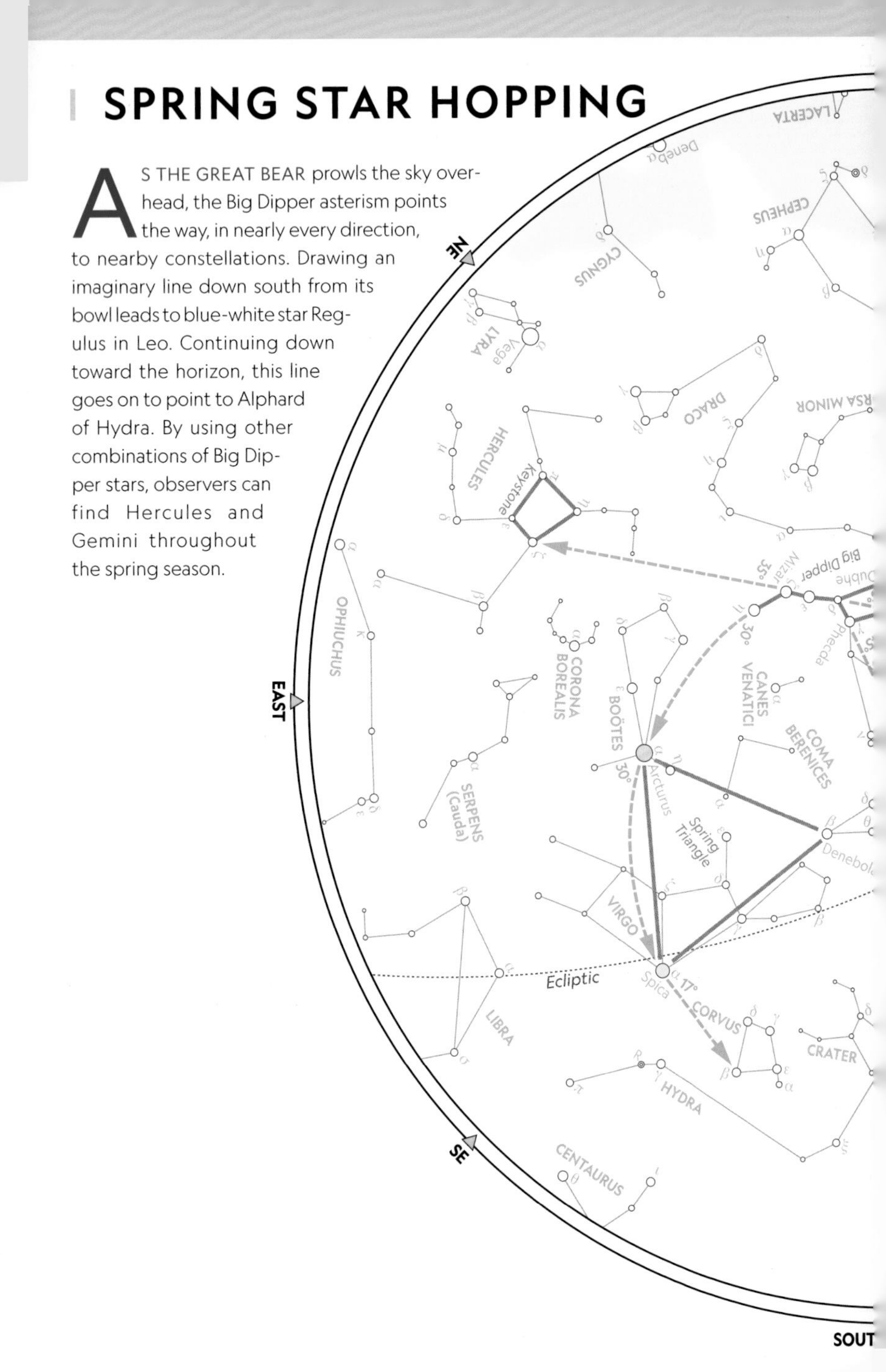

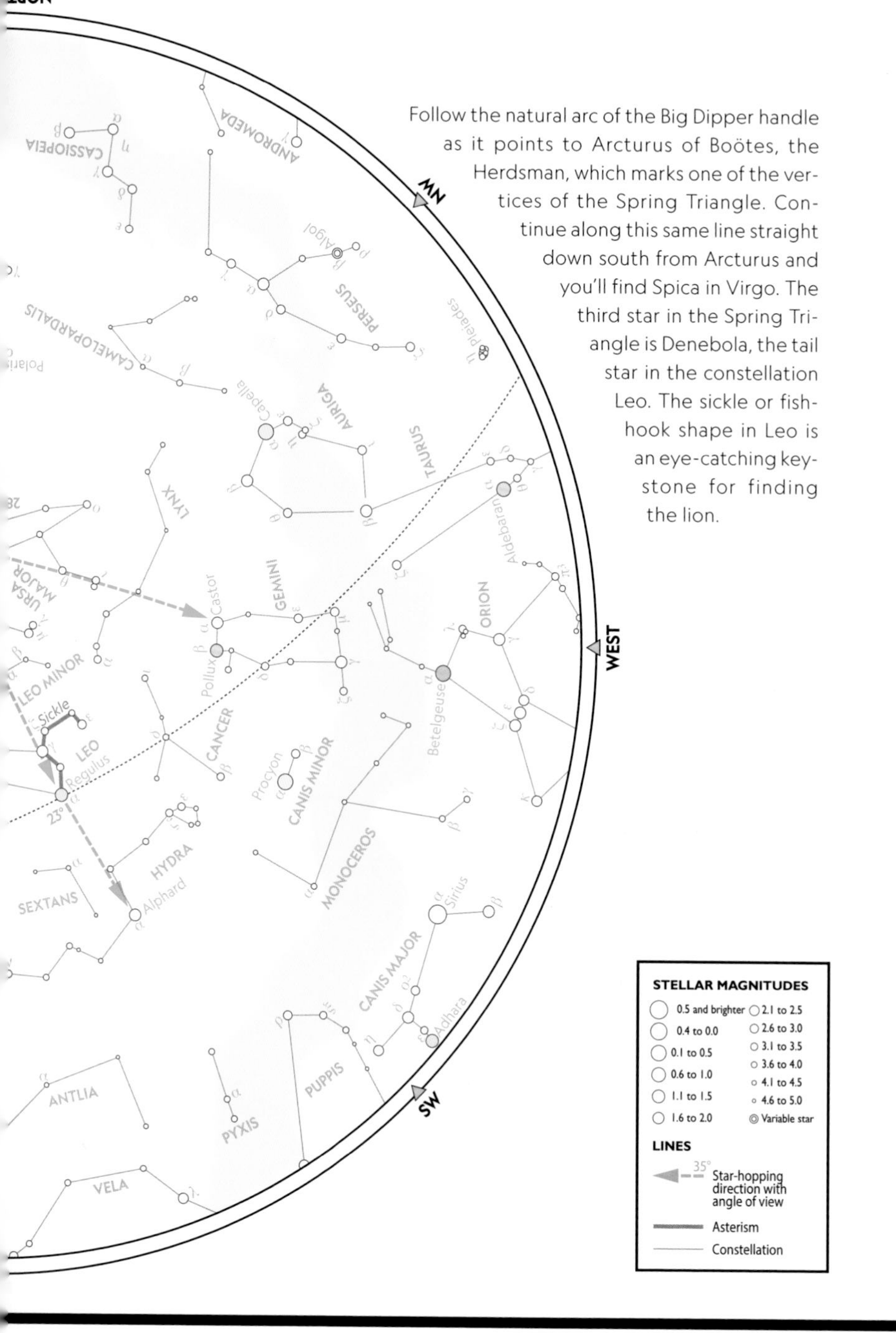

Follow the natural arc of the Big Dipper handle as it points to Arcturus of Boötes, the Herdsman, which marks one of the vertices of the Spring Triangle. Continue along this same line straight down south from Arcturus and you'll find Spica in Virgo. The third star in the Spring Triangle is Denebola, the tail star in the constellation Leo. The sickle or fishhook shape in Leo is an eye-catching keystone for finding the lion.

URSA MAJOR: The Great Bear

MAKEUP: 20 stars

BEST VIEWED: Mar./Apr.

LOCATION: Center of chart

DEEP-SKY OBJECT: Alcor & Mizar

FURTHER

The Big Dipper's stars, with the exception of Alkaid and Dubhe, are part of the same open star cluster and were likely formed in the same nebula. These stars are approximately 76 light-years away, making the Big Dipper the closest cluster to Earth.

URSA MAJOR, or the Great Bear, is one of the dominant shapes in the northern sky, including the seven-star asterism known as the Big Dipper (or the Plow in Britain). It is one of the most ancient constellations, associated with stories that have cut across many cultures and alternatively turned the Great Bear into a chariot, a horse and wagon, a team of oxen, and a hippopotamus (by Egyptians who had likely never seen a bear).

Stars & Objects

Its size and distinct shape make it a useful reference for locating other objects in the sky. From the north, over the course of the year, the Great Bear appears to run in a circle with its back to Polaris, the North Star. It's among the few constellations with a close to literal overall shape: The easily identifiable Big Dipper represents the rear torso and tail of the Bear, with the other stars mapping out its long nose and legs. The planetary nebula M97 or Owl Nebula is a very faint circular patch of light about 2.5 degrees southeast of

SIZE ON THE SKY:

INTERESTING OBJECTS in Ursa Major

A tour of Ursa Major begins with the stars Alcor and Mizar in the Dipper's handle. Visible as two stars to a sharp naked-eye observer, they are not actually bound in a binary star system. However, looked at through a small telescope, Mizar reveals itself as a true double, with Mizar B orbiting close by.

The bear also hosts a pair of galaxies 12 million light-years away—M81 and M82—that can be spotted together with binoculars. M81 is a spiral galaxy, while M82 is classified as a starburst galaxy with a high rate of star formation caused by an ancient collision with M81 over 100 million years ago.

Galaxies M81 and M82 bound by gravity

the Big Dipper star Merak. Shining at only magnitude 9.9, it has a very low surface brightness; so averted vision is particularly useful to bring out details when using smaller 4- to 6-inch (100 to 150 mm) telescopes.

Mythology

In Greek mythology, the Great Bear represents the nymph Callisto, who was transformed into a bear by an enraged Hera after discovering that her husband, Zeus, had impregnated her. Callisto's son, Arcas, mistakenly tried to kill her while hunting. Zeus intervened and placed them both in the sky.

Some Native American tribes believe the cup of the dipper represents a bear and the stars in the handle represent warriors that pursue it. In autumn, when the leaves turn, the constellation is low in the sky and the trees are thought to be stained red with the injured bear's blood.

MAIN STARS

ε | ALIOTH
COLOR: White
MAGNITUDE: 1.8
DISTANCE (LY): 81

α | DUBHE
COLOR: ORANGE
MAGNITUDE: 1.8
DISTANCE (LY): 124

η | ALKAID
COLOR: BLUE
MAGNITUDE: 1.9
DISTANCE (LY): 104

ζ | MIZAR
COLOR: BLUE
MAGNITUDE: 2.2
DISTANCE (LY): 78

β | MERAK
COLOR: WHITE
MAGNITUDE: 2.3
DISTANCE (LY): 79

URSA MINOR: The Small Bear

MAKEUP: 17 stars

BEST VIEWED: Year-round

LOCATION: Center-north

DEEP-SKY OBJECT: None

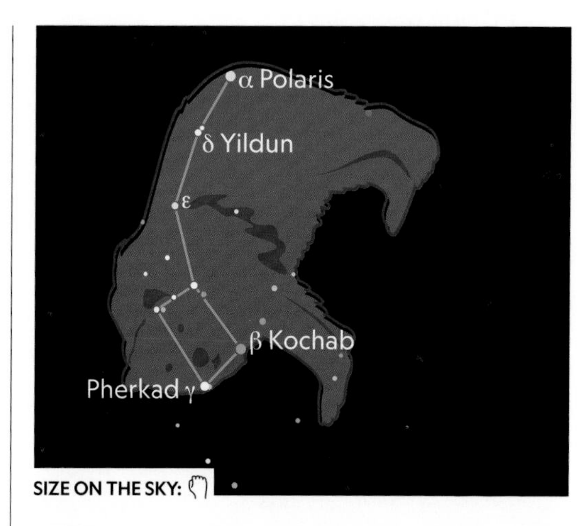

THIS NORTHERNMOST CONSTELLATION is visible year-round to star-watchers in the Northern Hemisphere. Its value as a navigational tool was recognized long ago, and travelers still look to the Little Dipper to locate Polaris. Other stars in the sky rotate around the celestial pole, making Polaris and the Little Dipper reliable constants in the night sky not only for travelers but also for those navigating the stars. Earlier cultures even used it as a clock, marking time as the constellation swung around the seemingly fixed point of the North Star.

Polaris is the most notable object in this small constellation. Through the recorded history of astronomy, the celestial pole has shifted. It is still approaching 433-light-years-distant Polaris and will reach its closest around the year 2100. The pole will continue to move past it and through a succession of new polestars, first Aldermain, then Deneb, and eventually Vega about 12,000 years from now. Telescopes 10 inch (250 mm) and larger begin to reveal structural details in the magnitude 11.2 barred spiral galaxy (NGC 6217). At 67 million light-years, the center nucleus can be spotted with smaller 4- to 6-inch (100 to 150 mm) telescopes.

Look for the annual Ursid meteor shower, which peaks in the predawn hours on December 21.

Barred spiral galaxy NGC 6217

BOÖTES: The Herdsman

MAKEUP: 8 stars

BEST VIEWED: June

LOCATION: Center of chart

DEEP-SKY OBJECT:
Izar, double star

Gaze in the direction of Boötes and you will find the emptiest part of space yet found, called the Great Void. Discovered in 1981, this mysterious expanse of relative emptiness stretches some 350 million light-years across and is 700 million light-years from Earth. Whereas such a great area would be expected to contain tens of thousands of galaxies, astronomers have found the Great Void contains only about 60 galaxies.

THIS ANCIENT CONSTELLATION is one of the most distinct in the late spring and early summer sky. Boötes is most easily identified by its brightest star, 36-light-year-distant Arcturus. It is simple to find this brilliant star not only because it is the third brightest in the sky but also because it is on an arc that connects to the handle of the Big Dipper—a reference point for the "arc to Arcturus." Train a 3-inch (75 mm) or larger telescope on Epsilon Boötes to split the close double star into the bright orange giant Izar and its pale white companion star.

The Quadrantid meteor shower falls from the northern part of the constellation and is one of the strongest of the year, producing several dozen meteors an hour. It takes place in the first week of January and originates at the meeting point of Boötes, Hercules, and Draco. Quadrantid meteors, however, are visible in substantial numbers only for a few hours.

Mythology

There are many different legends surrounding Boötes. According to some, he pursues the bears Ursa Major and Ursa Minor through the sky. In Greek, Arcturus means "Guardian of the Bear."

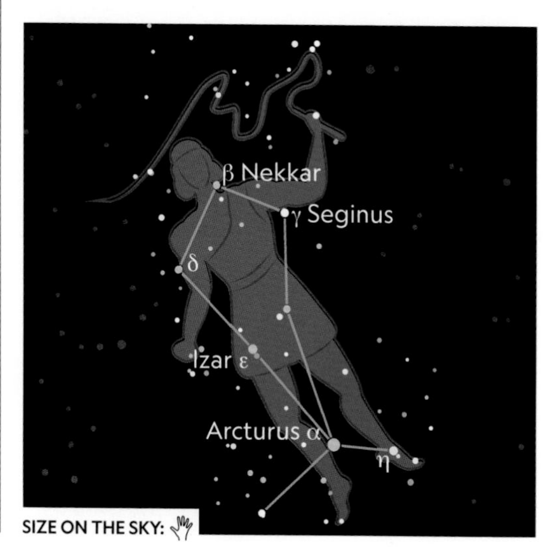

SIZE ON THE SKY:

LEO: The Lion

MAKEUP: 12 stars

BEST VIEWED: Mar./Apr.

LOCATION: Center of chart

DEEP-SKY OBJECT:
R Leonis, variable star

LEO IS A LARGE and easily recognized constellation that sits in a rather empty portion of sky, just beyond Ursa Major and between its fellow zodiacal constellations Cancer and Virgo. A sickle-shaped asterism represents the lion's head, mane, and chest. Shaped like a backward question mark, the "period" of this giant piece of celestial punctuation is marked by the brilliant star Regulus, while the stars trailing to the east mark Leo's hindquarters. This is one of the easier constellations to construct mentally, as it resembles the classic image of the Sphinx.

Regulus

The beacon-like, blue-white light of Regulus has long made it easy to see. Observation records of it date back to Babylonian tablets from about 2100 B.C. The sphinx, an icon of ancient Egyptian civilization, may have been modeled after this celestial beast. Better known in antiquity as Cor Leonis, the Lion's Heart, its current name is actually taken

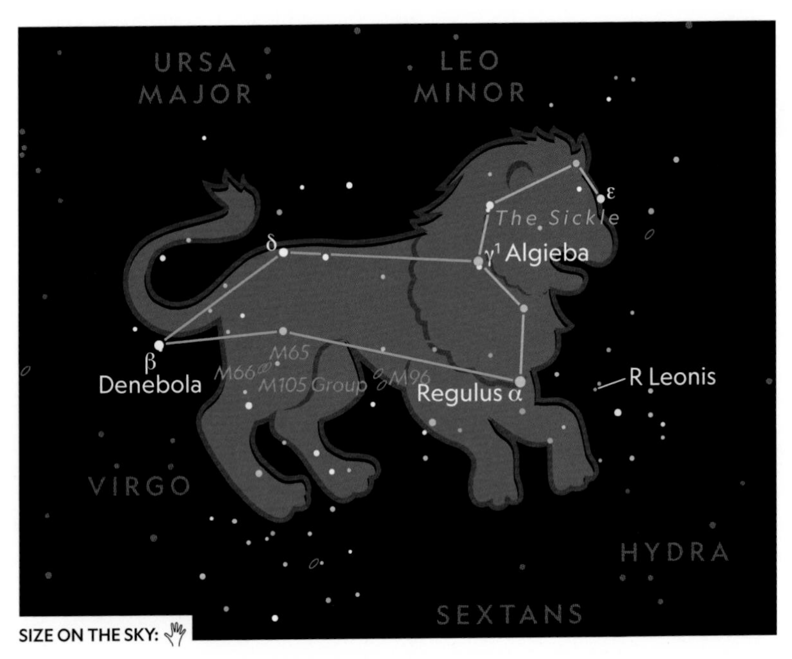

SIZE ON THE SKY:

Elliptical galaxy M105 from the Leo I Group

from a Latin word meaning "little king," and it is often known as the Royal star. This reference might be connected to Alexander the Great, who during his lifetime ruled the entire known world and was born during the Lion month.

Regulus lies 84 light-years away, shining some 160 times brighter than our own sun and with a diameter five times larger. Binoculars and small telescopes show a very dim companion star. The fainter companion's real distance from Regulus is about 100 times the distance tiny Pluto orbits from our sun.

Bright Stars & Objects

A group of galaxies including M65, M66, M95, M96, and M105 are visible just beyond the belly of the Lion but require at least a small telescope to detect. Close to Regulus is the stunningly fiery red variable star R Leonis, which cycles between magnitude 4.4 and 11.3 every 312 days. At its peak brightness it is faintly visible to the naked eye and an easy target for binoculars. Then as it sinks to its faintest, it becomes strictly a telescope target. R Leonis is so large that if it would replace our sun, its edge would reach out to the orbit of Jupiter.

INTERESTING OBJECTS in Leo

Algieba, meaning "lion's mane," is the second brightest star in the curve of the sickle and a giant: more than 10 times larger, and 80 times brighter, than our own sun. It is a beautiful example of a double star, with a companion that will glow a greenish yellow through binoculars. Lying over 100 light-years away, it takes more than 600 years for these two gravitationally locked exotic suns to circle each other..

MAIN STARS

α | REGULUS
COLOR: BLUE
MAGNITUDE: 1.36
DISTANCE (LY): 79

β | DENEBOLA
COLOR: BLUE
MAGNITUDE: 2.1
DISTANCE (LY): 36

δ | ZOSMA
COLOR: WHITE
MAGNITUDE: 2.6
DISTANCE (LY): 58

ε | ALGENUBI
COLOR: YELLOW
MAGNITUDE: 3.0
DISTANCE (LY): 250

γ^1 | ALGIEBA
COLOR: ORANGE
MAGNITUDE: 3.2
DISTANCE (LY): 130

VIRGO: The Virgin

MAKEUP: 13 stars

BEST VIEWED: May/June

LOCATION: Southeast

DEEP-SKY OBJECT:
M104, Sombrero galaxy

VIRGO IS ANOTHER of the zodiacal constellations that straddles the ecliptic band and the only one representing a woman. It was first cataloged by the Greek astronomer Ptolemy in the second century. The magnitude-1 star Spica can be located by star hopping: From the end of the handle of the Big Dipper, move south in the "arc to Arcturus" in the constellation Boötes, then "speed on to Spica" directly beneath it.

Stars & Objects

Located some 263 light-years from Earth, brilliant blue-white Spica is the fifth brightest star in the entire heavens: a blue giant about 14 times the mass of our own sun and 2,000 times more luminous. While it may look like a single star, it is in fact a double. Both are hot blue giants that orbit each other only 11 million miles (18 million km) apart. Spica lies along the path our moon takes across the sky, so at times it can be occulted (eclipsed) as it seems to disappear behind the moon for an hour or so. Virgo's second brightest star, Porrima, is a pretty double. Appearing faint to the naked eye, the two stars are slowly sepa-

FURTHER

Just east of Spica is the unique Sombrero galaxy (M104), marked by the distinctive dark dust lane that cuts across its middle and acts as the brim of the hat that provided the galaxy's name. Shining at magnitude 9, the Sombrero is one of the brightest spiral galaxies visible through backyard telescopes. Stretching some 130,000 light-years across, it is slightly larger than our Milky Way.

SIZE ON THE SKY:

INTERESTING OBJECTS in Virgo

Virgo cluster of galaxies

While Virgo may appear barren to the naked eye, and even binoculars, it is a galactic treasure trove for backyard observers. Spanning more than five degrees of the sky, an area about 10 times that of the full moon, the Virgo cluster is a vast collection of more than 2,000 spiral and elliptical galaxies huddled together in a 15 million light-year diameter. With its center about 53 million light-years away, the giant cluster is one of the closest such groups to our own Milky Way. The gravitational force of this cluster is so strong that it's slowly pulling our galaxy toward it. Through studying the movement of the galaxies, scientists realized that this region contains more dark matter than visible material.

rating from each other in the coming years as they orbit each other.

Virgo is among the richest areas of the sky to explore. It is prime territory to spot galaxies, and is the home of what is the most distant night sky object backyard observers are likely to see by themselves: quasar 3C273 Virgin. The magnitude-13 quasar is considered one of the closest of its kind to Earth at some three billion light-years away. A minimum 8-inch (200 mm) telescope is needed to spot it. It is estimated to be over four trillion times more luminous than our sun.

To the north, near the edge of the constellation's border, is the heart of what is known as the realm of galaxies, or the Virgo-Coma cluster, scattered between Virgo and nearby Coma Berenices. There are some 3,000 galaxies in this region, and with some casual trolling with 8-inch (200 mm), or even smaller, scopes with wide-field views, you can see dozens if not hundreds of the more prominent ones. The bright, fuzzy core of elliptical M49 is one of the easiest to spot, as is the twin sight of M84 and M86 in the same field of view. M87, a supergiant elliptical may never render more than a fuzzy view, but it's home to many trillions of stars and over 15,000 globular clusters. Professional telescopes have shown cosmic jets of superheated gas shooting out from the supermassive black hole at its core.

MAIN STARS

α | SPICA
COLOR: BLUE
MAGNITUDE: 0.1
DISTANCE (LY): 260

γ | PORRIMA
COLOR: WHITE
MAGNITUDE: 2.7
DISTANCE (LY): 38

ε | VINDEMIATRIX
COLOR: WHITE
MAGNITUDE: 2.9
DISTANCE (LY): 102

ζ | HEZE
COLOR: WHITE
MAGNITUDE: 3.4
DISTANCE (LY): 74

δ | AUVA
COLOR: RED
MAGNITUDE: 3.4
DISTANCE (LY): 202

CANCER: The Crab

MAKEUP: 5 stars

BEST VIEWED: Mar./Apr.

LOCATION: Southwest

DEEP-SKY OBJECT:
M44, Beehive cluster

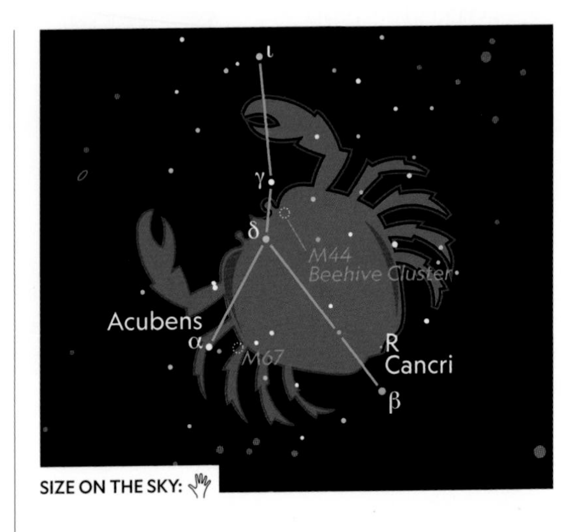

TINY CANCER, THE CRAB, lies between Gemini and Leo on the zodiac. In the sky, it can be found just northeast of Orion's belt. Despite being a member of the zodiac, Cancer is relatively faint, with no stars brighter than magnitude 4.

Deep-Sky Objects

Cancer includes one of the more interesting groups of stars: The Beehive cluster (M44). It is located just east of the Crab's head (the delta star in the center of the constellation), 577 light-years away. Visible to the naked eye under dark skies as a smudge of light, it was spotted as far back as the second century A.D. by stargazer Claudius Ptolemy. Binoculars bring 200-plus stars into view, appearing to span the same amount of sky as two full moons. In 2012, scientists found two new gas giants orbiting sunlike stars in M44, the first planets detected within a star cluster.

The 2,700 light-years-distant open cluster M67 contains 500 stars. At magnitude 6.1, it can be glimpsed with binoculars, but it really dazzles when seen through small scopes. The long-period variable star R Cancri is of interest as well; it fluctuates between magnitudes 6.2 and 11.2 in a year.

CANES VENATICI: The Hunting Dogs

MAKEUP: 2 stars

BEST VIEWED: May/June

LOCATION: Center of chart

DEEP-SKY OBJECT:
M51, Whirlpool galaxy

INTERESTING OBJECTS in Canes Venatici

At more than 60,000 light-years across and 31 million light-years away, the Whirlpool galaxy (M51) is one of the night sky's most beautiful spirals. Currently, M51 is interacting with NGC 5195, a dwarf galaxy whose gravitational disturbance is triggering star formation in the larger galaxy. A telescope with aperture measuring at least 6-inches (150 mm) wide will begin to reveal its spiral structure. You'll find it near the border of Canes Venatici, on a line traced from Cor Caroli to the final star in the Big Dipper's handle.

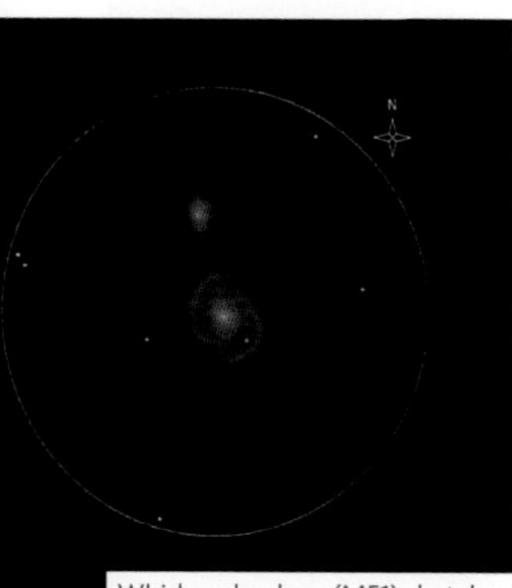

Whirlpool galaxy (M51) sketched as seen through a telescope

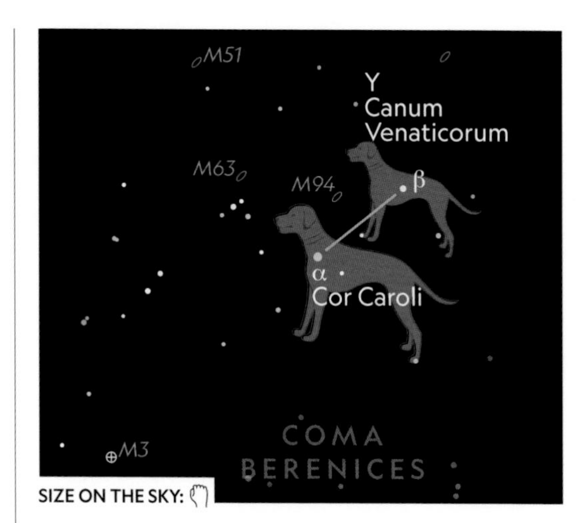

SIZE ON THE SKY:

THIS TWO-STAR CONSTELLATION can be found just below the handle of the Big Dipper—or, alternatively, running between the legs of Ursa Major, the bear that these dogs are chasing.

The two stars of Canes Venatici are surrounded by a multitude of interesting objects, some visible to the naked eye. Y Canum Venaticorum, or La Superba, is a vivid magnitude-5 red giant star that varies over the course of 157 days to magnitude 6.6. Cor Caroli—the Heart of Charles—is the brightest of the constellation, reputedly named after English King Charles II.

Deep-Sky Objects

Globular cluster (M3) can be found midway between Cor Caroli and Arcturus in Boötes. It is part of the famous catalog Charles Messier began in 1764, and it is of interest to astronomers since it hosts 274 variable stars—the most for any cluster ever seen. This stellar swarm is made of up to 500,000 stars, stretched across 200 light-years, and is located roughly 34,000 light-years from Earth. With a magnitude of 6.2, M3 can be spotted easily with binoculars, but a telescope will help resolve its outer members. An easy target for small telescopes is the magnitude-8.2 spiral galaxy M94.

HYDRA: The Sea Serpent

MAKEUP: 17 stars

BEST VIEWED: Mar./Apr.

LOCATION: Southwest

DEEP-SKY OBJECT: M83, spiral galaxy

FURTHER

In Greek myth, when Hercules battled the multiheaded serpent monster Hydra, it appeared invincible: Each time he chopped off one of its heads, more grew in its place. Hydra was only defeated when each stump was burned to prevent the growth of new heads.

THIS MASSIVE MONSTER OCCUPIES a larger strip of sky than any other constellation, winding its way from Libra to Cancer. A tight kite-shaped cluster, representing the serpent's head, leads its line of stars. Alphard, or the Solitary One, is Hydra's brightest star and can be located by looking just east of the line between Regulus in Leo to Sirius in Canis Major. With a tail that drops close to the southern horizon from mid-northern latitudes, Hydra does not match its size with luminosity: Save for two stars, its dispersed members are all magnitude 3 or dimmer. While technically a Southern Hemisphere constellation, it can be seen as far as the 54-degree latitude line north of the Equator.

Stars & Deep-Sky Objects

Spiral galaxy M83 at magnitude 8 makes for a beautiful telescope target for deep southern observers. Open cluster M48, shining at magnitude 5.5, is probably the easiest deep-sky object for northern latitude observers to hunt down with binoculars. The Mira star R Hydrae has been recognized as a variable since the late 17th century. Watch it over the course of 390 days as its luminosity varies from a naked-eye magnitude 3.5 to a dim binocular 10.9. V Hydrae is worth hunting for its color—a lush red.

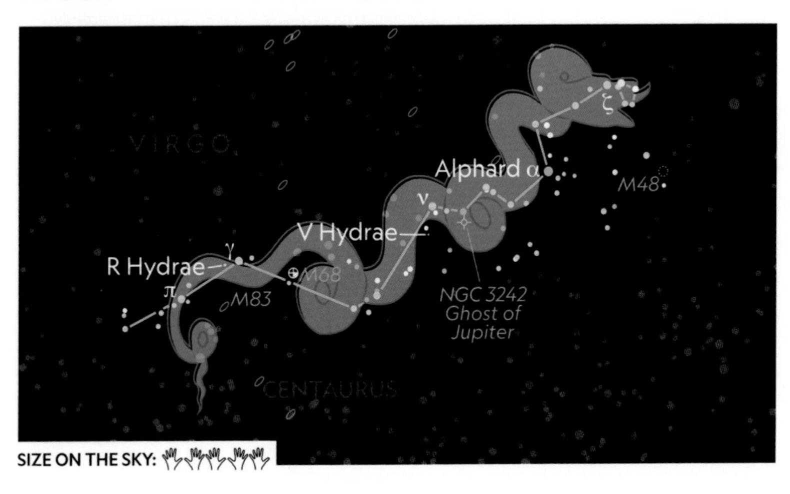

SIZE ON THE SKY:

COMA BERENICES: Berenice's Hair

MAKEUP: 3 stars

BEST VIEWED: May/June

LOCATION: Southeast

DEEP-SKY OBJECT:
M64, Black Eye galaxy

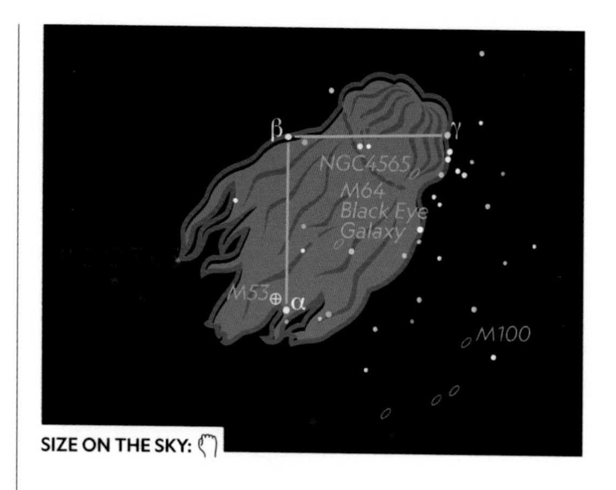

SIZE ON THE SKY:

COMA BERENICES WAS ONCE part of nearby constellations—tucked into Virgo as a wisp of her hair, and considered a tuft of Leo's tail. In the 1500s cartographers and astronomers, including Caspar Vopel and Tycho Brahe, separated this small-but-rich area of the sky from the others and rejuvenated the tale of an Egyptian queen to name it.

Coma Berenices is located just north of Virgo and south of Canes Venatici and the Big Dipper. Its proximity to the Virgo cluster of galaxies makes for interesting skywatching. The Black Eye galaxy (M64) is easily visible by telescope in between the two outermost stars in Coma Berenices—effectively at the base of a triangle. With a 4- to 6-inch (100 to 150 mm) telescope, a dark cloud of dust in front of its core creates the illusion of a "black eye" in the middle of this spiral galaxy. Fine globular cluster M53 is also easily found next to the constellation's brightest star, alpha Comae Berenices.

Mythology

Berenice was a queen of Egypt during the reign of Ptolemy III. As he went to war, his wife struck a deal with the goddess Aphrodite, promising her long, beautiful hair in return for the safe return of her husband. An astronomer of the royal court convinced the ruling couple that a grateful Aphrodite had placed the queen's gift in the stars.

INTERESTING OBJECTS in Coma Berenices

One of the brightest and largest spiral galaxies in the Coma cluster is 55-million-light-years-distant M100. While appearing as a faint magnitude-9 patch of light in small telescopes, it is most famous for hosting five visible supernovae explosions in the past century.

CORONA BOREALIS: The Northern Crown

MAKEUP: 7 stars

BEST VIEWED: June/July

LOCATION: Center of chart

DEEP-SKY OBJECT:
T Coronae Borealis

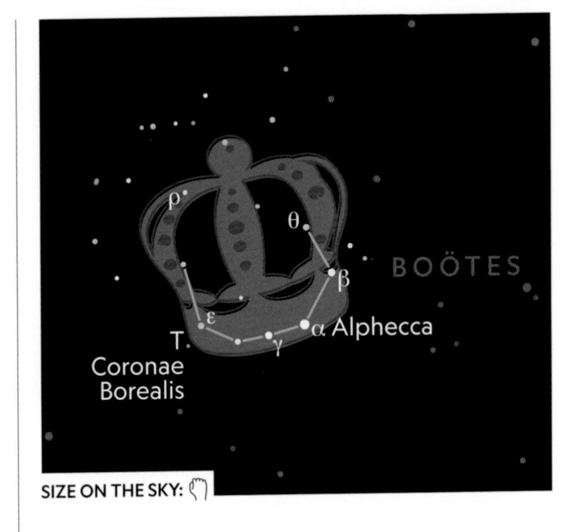

A N OBVIOUS CROWNLIKE SEMICIRCLE shape compensates for the small size and faint magnitude of this constellation squeezed between Boötes and Hercules. You can locate it just south of a line traced between the bright stars Vega in Lyra and Arcturus in Boötes.

One of the constellation's stars varies greatly in its brightness. In 1866, and again in 1946, T Coronae Borealis rose from magnitude 10 to magnitude 2, and then faded back, as the result of a nova explosion. When this eruption inevitably occurs again, this star—which, at magnitude 10, is currently invisible to the naked eye—will become the brightest one in the constellation.

Mythology

The Northern Crown is primarily associated with Greek myth and legend, where the crown belongs to Ariadne, the daughter of the King of Crete. Reluctant to accept a marriage proposal from the god Dionysus, she asked him to prove his powers. He threw the crown into the heavens as a tribute to her, which did the trick. Alphecca is the Arabic name for the alpha star, but another proper name is Gemma, which means "gem" or "jewel" in Latin.

Rho Coronae Borealis (ρ) is under investigation for extrasolar planets. One planet larger than Jupiter has been found, and astronomers suspect it's not alone. At random intervals, as dark material erupts in its atmosphere, this variable star's magnitude drops sharply from 5.8 to 14.8.

LEO MINOR: The Small Lion

MAKEUP: 3 stars

BEST VIEWED: Mar./Apr.

LOCATION: Center of chart

DEEP-SKY OBJECT:
R Leonis Minoris, variable star

NGC 3432
β
46
21
R Leonis Minoris
NGC 3486
NGC 3344
LEO

SIZE ON THE SKY:

THIS GROUP is one of the more recent constellations, designated by Johannes Hevelius in the 17th century. The small, modest grouping requires a dark night to be seen at all, as no star in the constellation is brighter than 46 Leonis Minoris at magnitude 3.8. In mid-northern latitudes, it can be seen almost directly overhead during March and April, perched between Leo and Ursa Major.

Stars & Objects

The most notable of the Small Lion's dim stars is R Leonis Minoris, east of the three stars that form the constellation. It is a Mira-like red giant, long-period variable star that will oscillate over a year between magnitude 6.3 and 13.2, the limit of a 6-inch (150 mm) scope. Its brightest deep-sky objects include the magnitude-10 spiral galaxies NGC 3486 and NGC 3344, located 33 million and 25 million light-years away, respectively. In moderately sized scopes of 8 to 10 inches (200 to 250 mm) both show fairly bright, but diffuse halos surrounding bright cores. Another fine target for backyard scopes is the magnitude-11 barred spiral (NGC 3432), also known as the Knitting Needle galaxy, which takes its name from its very slender, edge-on appearance.

NGC 3486

CORVUS: The Crow

MAKEUP: 5 stars

BEST VIEWED: Apr./May

LOCATION: Southeast

DEEP-SKY OBJECT: NGC 4038 and NGC 4039, Antennae galaxies

CORVUS AND CRATER ARE BOTH associated in myth with Hydra, the giant nearby serpent. The Crow is located near Hydra's tail and just west of Spica in nearby Virgo. Alchiba, the constellation's alpha star, is just outside the polygon of the bird's body, representing its downward-facing head.

Just outside of Corvus, the Antennae is a pair of colliding galaxies (NGC 4038 and NGC 4039) also known as the Ring-tailed galaxy. The pair can be seen with an 8-inch (200 mm) telescope, with the "tail" near the border of Corvus and Crater.

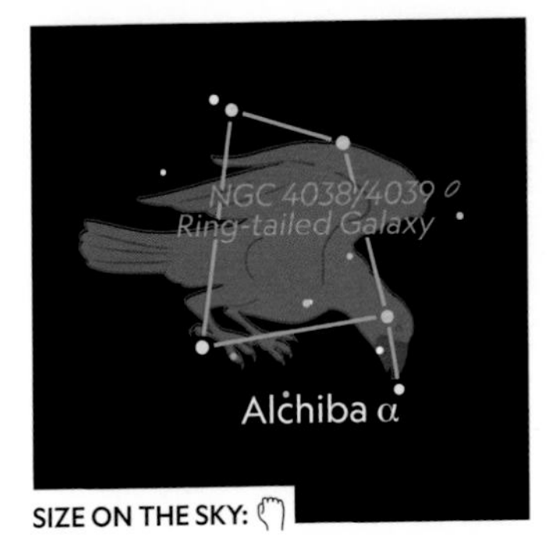

SIZE ON THE SKY:

Mythology

The Greek god Apollo sent a crow for a cup of water, but the bird got distracted by a tasty-looking fig. Though it eventually returned to Apollo, it was banished into the heavens for its tardiness.

CRATER: The Cup

MAKEUP: 8 stars

BEST VIEWED: Apr./May

LOCATION: Southeast

DEEP-SKY OBJECT: NGC 3887, galaxy

THIS IS A SMALL CONSTELLATION that has been interpreted to take on many forms. To some, Crater first represented a spike on the back of the sea monster Hydra. Keeping that in mind will help to locate this dim constellation. From the Northern Hemisphere, Crater appears deep in the southern sky during the mid- to late spring, right above Hydra and just west of Corvus. Locating Spica in Virgo may also help, as both Crater and Corvus are southwest of the bright star. The shape of the cup is unmistakable: Four stars establish its base, while four more form a goblet opening toward Spica. There are few bright deep-sky objects in Crater, save for dim galaxies of magnitude 11 or less such as NGC 3887.

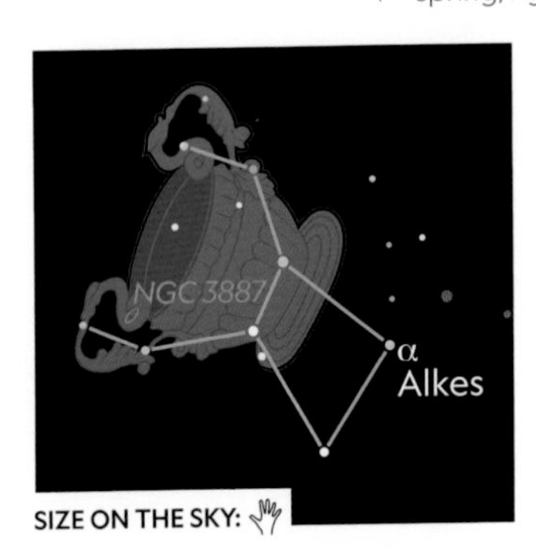

SIZE ON THE SKY:

Mythology

Most early observers of this constellation saw it as a cup. In Greek mythology, Crater represents the vessel brought to Apollo by Corvus.

ANTLIA: The Air Pump

MAKEUP: 4 stars

BEST VIEWED: Mar./Apr.

LOCATION: Southwest

DEEP-SKY OBJECT: NGC 2997, spiral galaxy

FORMERLY NAMED ANTLIA PNEUMATICA, this primarily southern constellation can be spied in the springtime by stargazers in the Northern Hemisphere when looking toward the southern horizon. It lies under the sprawling form of Hydra, about five fist-widths south of the star Regulus in Leo. Antlia's alpha star is its brightest one, but it hasn't been given a proper name.

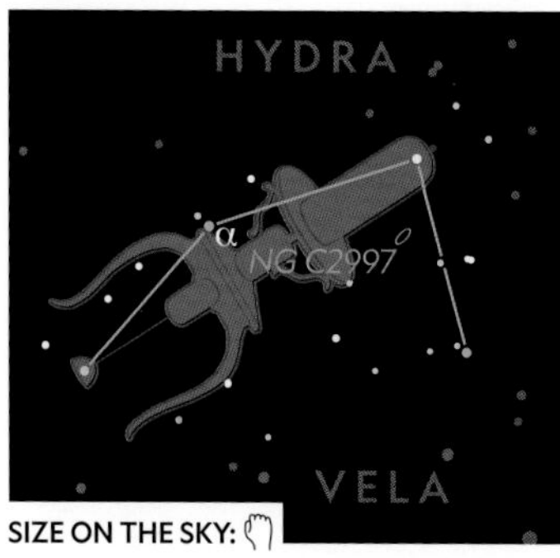

SIZE ON THE SKY:

Just inside the constellation's corner is a distant spiral galaxy, Antlia's deep-sky object: NGC 2997, 55 million light-years away. To an amateur astronomer, this large, faint galaxy, though difficult to see, will appear through a small telescope as an oval fuzzy spot with a dark ring surrounding its center. Powerful 16-inch (400 mm) scopes will reveal the galaxy's tight spiral arms coiled counterclockwise, each peppered with red, knotlike star formation regions.

SEXTANS: The Sextant

MAKEUP: 3 stars

BEST VIEWED: Mar./Apr.

LOCATION: Southwest

DEEP-SKY OBJECT: NGC 3115, Spindle galaxy

SEXTANS URANIAE'S FULL NAME has been shortened over the years to just Sextans. This small, faint constellation is the 17th-century creation of astronomer Johannes Hevelius and fills out the sky near Leo's portion of the ecliptic. It is composed of a handful of stars, none brighter than magnitude 4.5.

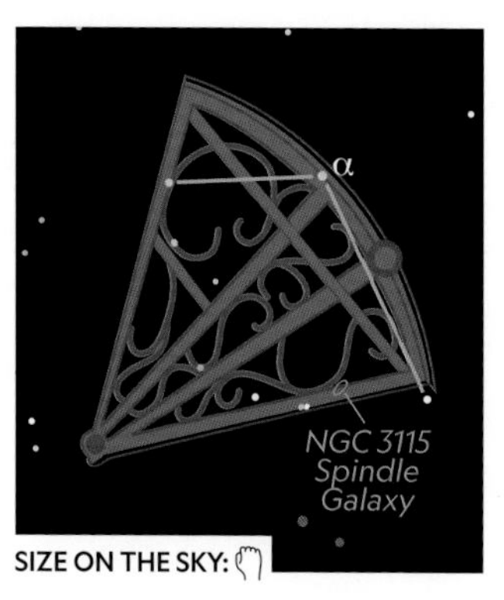

SIZE ON THE SKY:

Sextans can be found in the south, occupying the space between Hydra and Leo. Its small patch of sky contains one notable deep-sky object—the Spindle galaxy (NGC 3115), viewed edge-on as a magnitude-9 flattened disk. Visible through even binoculars on dark, moonless nights with no light pollution, this lens-shaped galaxy lies 32 million light-years away. Many times larger than our Milky Way, it hides at its core one of the largest, closest supermassive black holes ever discovered beyond our galaxy, which weighs in at a whopping two billion times the mass of our sun.

SUMMER

THE ORANGE STAR ARCTURUS pins down the constellation of Boötes. Next door is the distinctly curved constellation of Corona Borealis and the mythical strongman Hercules riding high in the sky. Lyra's main star, Vega, will be almost directly overhead, unmistakably bright at magnitude 0. Crossing the Milky Way to the east is the Summer Triangle asterism marked by three of the season's brightest stars and their respective constellations: Lyra, Cygnus, and Aquila.

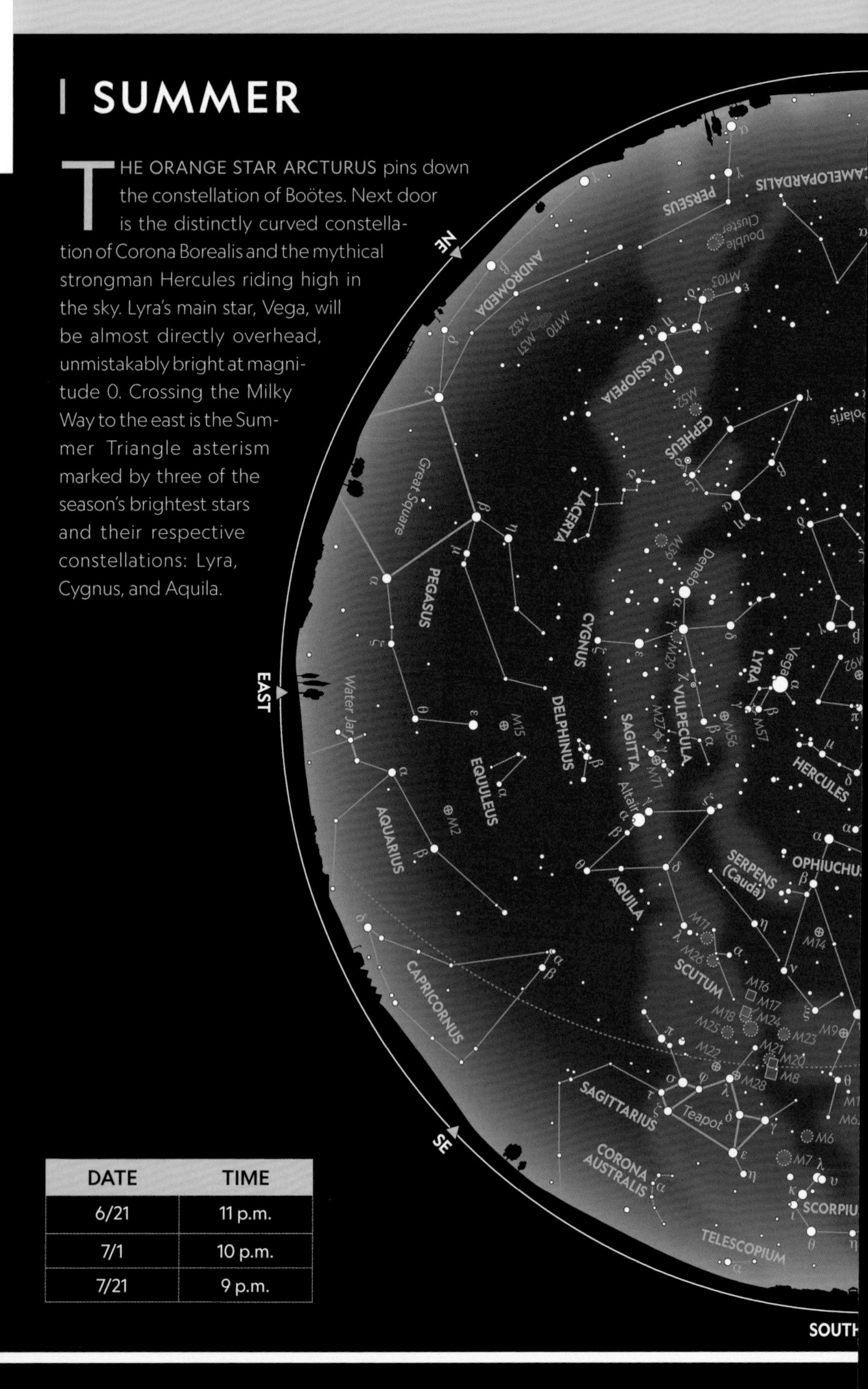

DATE	TIME
6/21	11 p.m.
7/1	10 p.m.
7/21	9 p.m.

The high southern sky is dominated by the sweeping constellation of Ophiuchus entangled with the Serpens (the Serpent), which together occupy approximately three hand-widths of the sky. The bright orange star Antares marks the southerly constellation Scorpius. Sagittarius points the way to the core of the Milky Way and is an area rich in nebulae and star clusters, like the Lagoon (M8) and Trifid Nebulae (M20), and the Great Sagittarius star cluster (M22).

STELLAR MAGNITUDES

- −0.5 and brighter
- −0.4 to 0.0
- 0.1 to 0.5
- 0.6 to 1.0
- 1.1 to 1.5
- 1.6 to 2.0
- 2.1 to 2.5
- 2.6 to 3.0
- 3.1 to 3.5
- 3.6 to 4.0
- 4.1 to 4.5
- 4.6 to 5.0
- Variable star

DEEP-SKY OBJECTS

- Open star cluster
- Globular star cluster
- Bright nebula
- Planetary nebula
- Galaxy

SUMMER STAR HOPPING

RIDING HIGH in the southeast sky is the season's biggest asterism, the Summer Triangle, with each of its corners marking the starting point to a constellation. The second brightest star in this stellar pattern, Deneb, marks the tail of Cygnus, the swan constellation also known as the Northern Cross. The Milky Way appears to run down the spine of the swan, into the constellations of Aquila and Scutum. Extend an imaginary line from the mythical bird, down to the southern horizon directly to the Teapot asterism within Sagittarius.

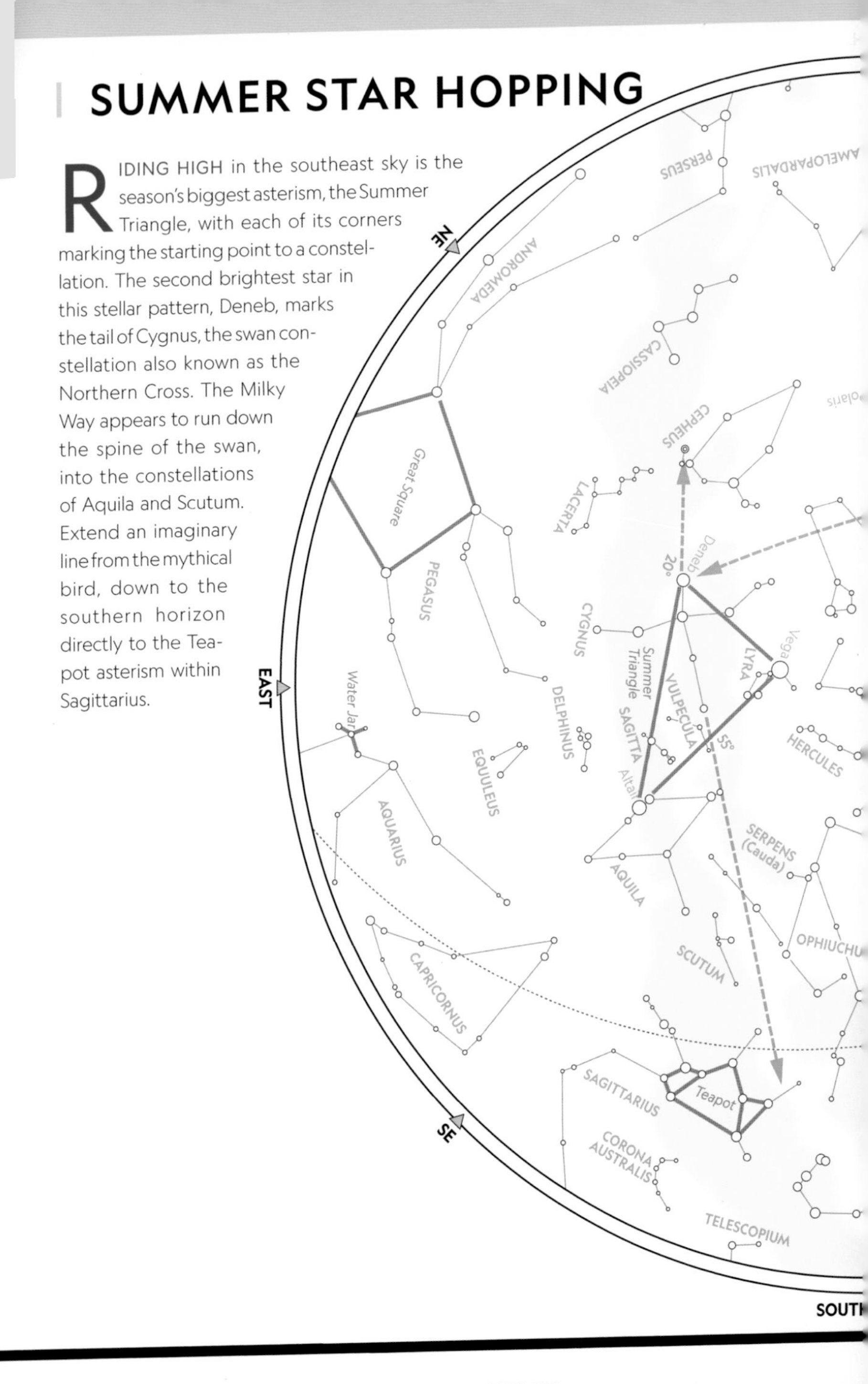

Hanging low in the northwest is the Big Dipper asterism. By continuing a line out from the handle stars to the next brightest stellar pattern, we reach the Keystone asterism, which marks the mythical hero constellation of Hercules. Extend down south the right side of the Keystone and we reach the bright star Antares, marking the heart of Scorpius. On moonless nights sweep this entire region from Cygnus to Scorpius with binoculars and cruise the countless star clouds and deep-sky treasures.

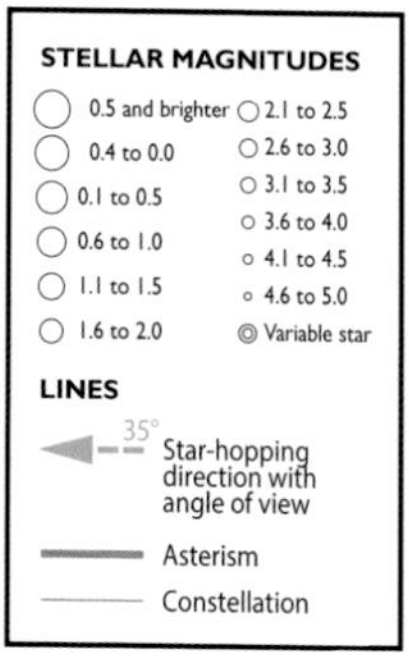

CYGNUS: The Swan

MAKEUP: 13 stars

BEST VIEWED: Aug./Sept.

LOCATION: Northeast

DEEP-SKY OBJECT: NGC 7000, North America Nebula

FURTHER

A thick band of dust lies within Cygnus, creating a dark space in the Milky Way that is easily visible to the naked eye under good conditions. This part of the Milky Way is known as the Cygnus Rift or the Northern Coalsack, a name borrowed from a similar dark spot found in the Southern Cross.

CYGNUS LIES in what for observers is a dense and fascinating part of the sky. The bird's wings span the Milky Way at a location packed with stars and an assortment of deep-sky objects. Owing to the positions of its bright stars it is sometimes called the Northern Cross—the northern parallel to the brilliant Southern Cross constellation in the Southern Hemisphere. You will see Cygnus highest in the sky in late summer and early fall, with its head pointing south like a bird on its migratory path toward warmer climates.

Stars & Objects

Deneb, the alpha star and tail of the bird, is a brilliant magnitude-1 star that joins Altair and Vega to form the Summer Triangle. Deneb is 25 times more massive than our sun and 60,000 times more luminous. Straddling the galactic equator is Cygnus's bright blue-and-gold binary beta star, Albireo. The split between its two stars is visible with even a small telescope.

The North America Nebula is a large, diffuse emission nebula 2,200 light-years from Earth, which appears as a

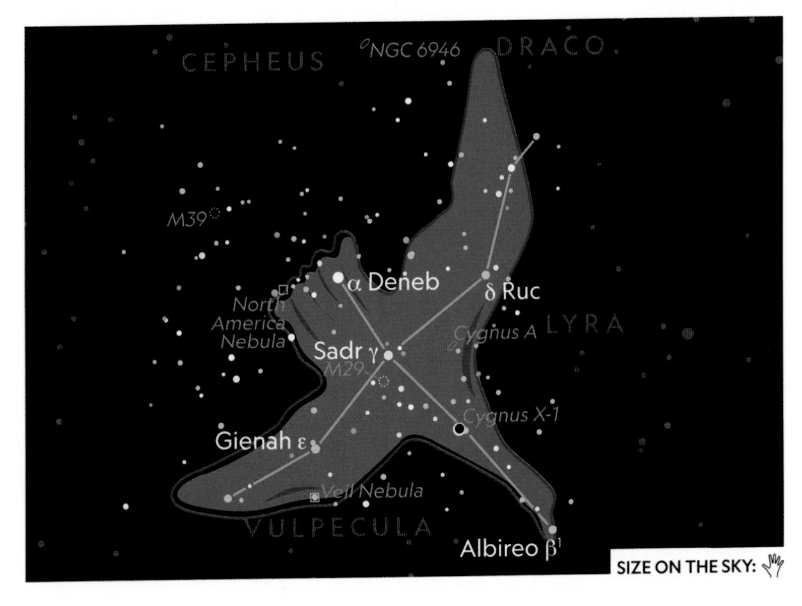

INTERESTING OBJECTS in Cygnus

The Veil Nebula, near Epsilon Cygni, is a gas cloud left over from a supernova explosion, and is best viewed using a high-contrast filter. M39 is a loosely bound open cluster that is visible through binoculars, but it was apparently first seen with the naked eye by Aristotle around 325 B.C. Cygnus X-1 was discovered in the mid-1970s. It was the first black hole discovered by observing its x-ray emissions, created by gas streaming in from a nearby star. The Cygnus A galaxy also centers on a supermassive black hole.

North America Nebula

colorful gas cloud on the Swan's southern border. You can see it through binoculars under excellent conditions, while a cloud of stars across its face is faintly visible to the naked eye. The nebula's signature resemblance to the North American continent appears most notably in photographs.

Mythology

In Greco-Roman mythology, Zeus transformed himself into a swan to seduce the maiden Leda, who later gave birth to the Gemini twins Castor and Pollux. The mythical musician Orpheus, son of Apollo, was thought to be transformed into a swan to be next to his beloved lyre—the nearby constellation Lyra. The constellation is also considered one of the Stymphalian birds, which Hercules hunted as part of his famous 12 labors.

MAIN STARS

α | DENEB
COLOR: White
MAGNITUDE: 1.3
DISTANCE (LY): 2,600

γ | SADR
COLOR: Yellow-white
MAGNITUDE: 2.2
DISTANCE (LY): 1,500

ε | GIENAH
COLOR: Yellow-orange
MAGNITUDE: 2.5
DISTANCE (LY): 72

δ | RUC
COLOR: Blue-white
MAGNITUDE: 2.9
DISTANCE (LY): 171

β^1 | ALBIREO
COLOR: Blue-yellow
MAGNITUDE: 3.3
DISTANCE (LY): 380

LYRA: The Lyre

MAKEUP: 5 stars

BEST VIEWED: July/Aug.

LOCATION: Center of chart

DEEP-SKY OBJECT:
M57, Ring Nebula

RESEMBLING A SMALL JAGGED LINE hitched to a parallelogram, Lyra is one of the delights that is easy to envision and easy to spot. Its alpha star, Vega, is among the brightest stars in the sky and is directly overhead from mid-northern latitudes in summer—a handy reference point for many other constellations and objects. With Deneb and Altair, Vega forms the Summer Triangle asterism.

Bright Stars

Three times the size of our sun and 50 times more luminous, astronomers think Vega may be young and newly formed, only about 400 million years old—that's less than 10 percent of our sun's age. A big Vegan discovery came in 1983, when the Infra-Red Astronomical Satellite (IRAS) photographed a disk of cool dust surrounding the star. Astronomers suspect that planets may be forming out of this swirling material around Vega, creating the birth of an alien solar system in the same way that ours formed.

Those with keen eyesight will notice a star that is actually an eye-catching pair of stars known as Epsilon Lyrae (ε). Orbiting each other around their common center of

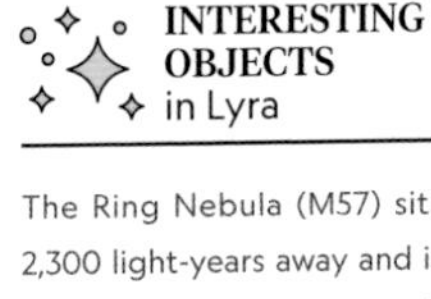

INTERESTING OBJECTS in Lyra

The Ring Nebula (M57) sits 2,300 light-years away and is visible with at least a 3-inch (75 mm) telescope with a high-powered eyepiece to reveal its donut shape. This famous ring is formed by gas that was emitted by a dying star. The nebula looks circular from our perspective, but scientists have concluded that it is actually cylindrical. Large backyard scopes may reveal a faint star in the center of this celestial smoke ring. Known as a white dwarf, it is the naked white-hot core of the exploded star. The same fate awaits the sun when it dies.

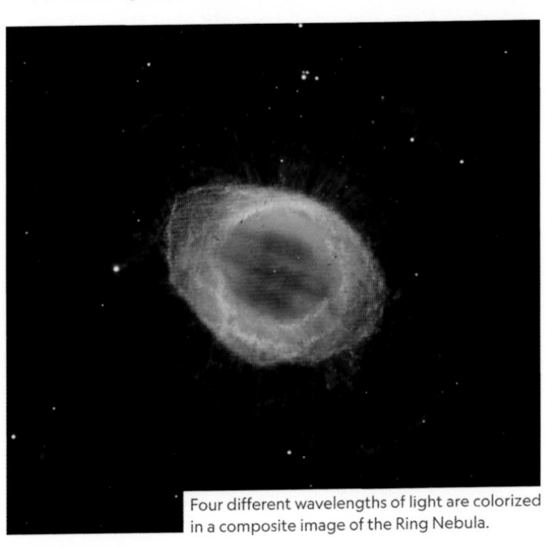

Four different wavelengths of light are colorized in a composite image of the Ring Nebula.

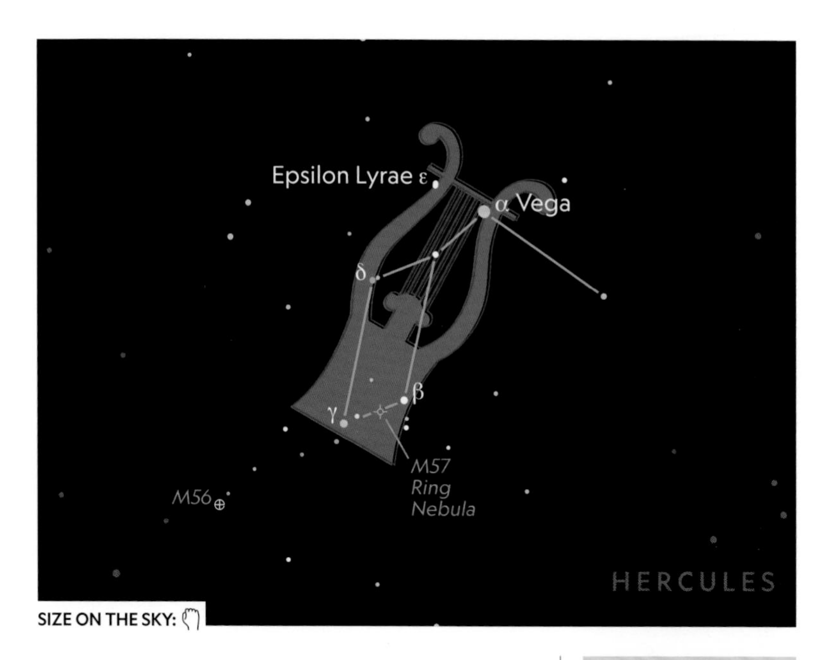

SIZE ON THE SKY:

gravity, they sit 175 light-years from Earth. Looking through a small telescope, observers will see that these celestial twins are actually both doubles themselves, forming a spectacular quadruple star system known popularly as the "double-double."

Mythology

A chilling love story accompanies this constellation. Apollo gave his son Orpheus the lyre and taught him to play captivating music. Despite being the object of many women's affections, Orpheus loved his wife Eurydice. She died and was sent to the underworld, but Orpheus was unyielding in his desire to bring her back to life. The gods decided to allow it, but Orpheus failed to follow their one admonition not to turn around to look back at his wife as they exited Hades. Having lost his love again, Orpheus refused all other advances and was killed by a discouraged group of young women. He was then reunited with his wife, and as a tribute to their love, Zeus sent the lyre up into the sky.

MAIN STARS

α | VEGA
COLOR: WHITE
MAGNITUDE: 0.0
DISTANCE (LY): 25

γ | SULAFAT
COLOR: BLUE-WHITE
MAGNITUDE: 3.3
DISTANCE (LY): 620

β | SHELIAK
COLOR: WHITE
MAGNITUDE: 3.5
DISTANCE (LY): 950

δ | DELTA LYRAE
COLOR: ORANGE-RED
MAGNITUDE: 4.3
DISTANCE (LY): 740

AQUILA: The Eagle

MAKEUP: 10 stars

BEST VIEWED: Aug./Sept.

LOCATION: Southeast

DEEP-SKY OBJECT:
Eta Aquilae, variable star

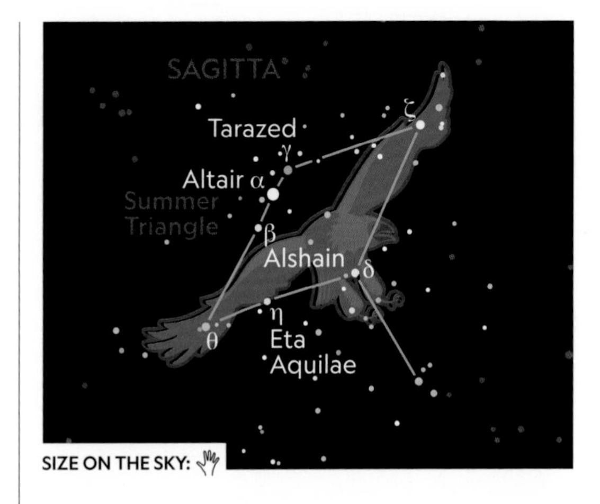

SIZE ON THE SKY: 

AQUILA, NAMED BY STARGAZERS in ancient Mesopotamia, flies close enough to Earth's Equator that it can be seen from anywhere in the world. But it is most easily found in the Northern Hemisphere when looking south in the middle of summer.

At nearly 17 light-years from Earth, Altair (Arabic for "the bird" or "the eagle") is the constellation's brightest star. It is also part of the trio of stars that forms the Summer Triangle asterism and is one of the brightest stars in the sky. The supergiant Eta Aquilae (η) is one of the easiest Cepheids to distinguish with the naked eye as it varies from magnitude 4.1 to 5.3 every 7.2 days, earning it the Hebrew name *bezek* meaning "lightning."

Bluish alpha star Altair with orange-tinted Tarazed

Mythology

The eagle belonged to Zeus and, according to myth, carried the god's thunderbolts for him, perhaps a reference to the blinking variable star. He was also said to have brought a young shepherd named Ganymede into the sky, who was to serve as Zeus's cupbearer. The youth would be immortalized as the nearby constellation Aquarius. Arab astronomers referred to the star Zeta Aquilae (ζ) as the eagle's tail, but in modern diagrams, the star represents one of the bird's wings.

CEPHEUS: The King

MAKEUP: 10 stars

BEST VIEWED: Sept./Oct.

LOCATION: Center of chart

DEEP-SKY OBJECT:
Delta Cephei, variable star

A CIRCUMPOLAR CONSTELLATION, Cepheus is visible all year in the Northern Hemisphere. Five bright stars make up the body of the king. They are in the shape of a small house, with a roof that points approximately toward Polaris. Errai, Cepheus's gamma star 45 light-years distant, is both a binary star and a host to an orbiting planet. Estimated at about 1.59 times the size of Jupiter, Errai's planet, named Tadmor, offers evidence that planets can form in relatively close binary systems. It's also one of the first exoplanets to be given an official name through a public nomination and voting process. Another star of interest is Mu Cephei (μ)—a stunning red "garnet star" about 5,000 light-years distant with an irregular period of variability.

Mythology

Cepheus was the king of Ethiopia and a luckless character in the myth of Perseus and Andromeda. His queen vainly boasted that their daughter was more beautiful than the daughters of Nereus, causing the enraged Poseidon to release a sea monster. An oracle told the king that he had to sacrifice his daughter Andromeda to appease the offended gods.

The star Delta Cephei (δ), at 887 light-years away, is considered the perfect model of a Cepheid variable and has about a 1-magnitude change of brightness, changing from 3.51 to 4.34 over the course of about 5.5 days. Cepheids with longer periods of variation have greater intrinsic brightness. So when we use a Cepheid's variation period to determine its true brightness and compare this with its apparent brightness, the Cepheid can help pin down the star's distance.

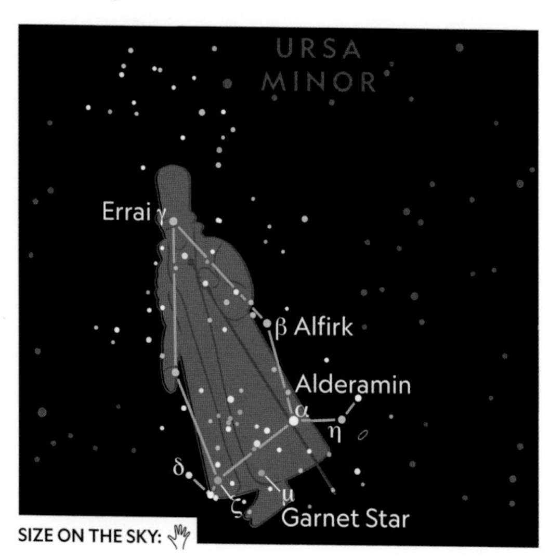

SIZE ON THE SKY:

HERCULES: The Strongman

MAKEUP: 20 stars

BEST VIEWED: July/Aug.

LOCATION: Center of chart

DEEP-SKY OBJECT:
M13, Hercules cluster

As one of the more prominent heroes of Greek mythology, Hercules has earned a central place in the sky—at least for those in the Northern Hemisphere—appearing almost directly overhead during summer months.

From mid-northern latitudes the warrior can be seen running east toward the Milky Way. Four stars in an approximate square make up the asterism known as the Keystone, similar in shape to the central stone in archways. Right next to bright Vega, Hercules is easy to locate, especially during the summer weeks when the constellation is at its highest. Though prominently positioned in the night sky, Hercules is a dim constellation that does not benefit from any notably bright stars and was once named "the Phantom." Only three are brighter than magnitude 3. Rasalgethi, the alpha star, is a red supergiant that is visible to the naked eye. As an exercise in distinguishing the colors of stars in a telescope, try splitting Rasalgethi from its blue-green binary companion.

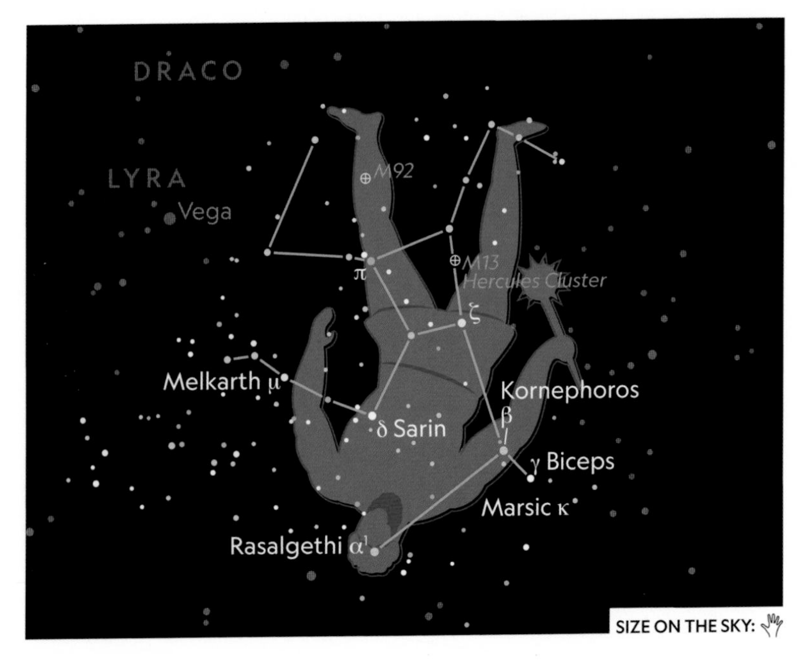

INTERESTING OBJECTS in Hercules

Hercules cluster

Considered the brightest globular star cluster visible in the northern sky, the Hercules cluster (M13) is 22,000 light-years away and home to over a million stars. Found on one side of the Keystone asterism, it shines at magnitude 5.8—just barely visible to the naked eye (and from a very dark location) as a circular blur. A 4-inch (100 mm) telescope will begin to show it as a tight ball of individual stars about 100 light-years across. Larger scopes will resolve its bright and densely packed core. A second, slightly more compact globular cluster, M92, also sits near the Keystone asterism and is easy to spot with binoculars even from suburbs.

Mythology

In Greco-Roman lore, Hercules was a half-mortal son of Zeus, famous for his amazing strength. He was driven mad by Hera, queen of the gods and Zeus's wife, who was angered by her husband's unfaithfulness. While cursed with insanity, Hercules took his family's life. When the tragic spell lifted, Hercules was so overcome with grief that he undertook 12 labors to repent for his actions. Each challenge seemed impossible, but through strength and ingenuity Hercules overcame each one and was granted celestial immortality. Many of the creatures Hercules battles are the other constellations in the sky including Leo the lion, Cancer the crab, and Hydra the serpent.

Other cultures assigned figures of great prominence to this constellation, and it has borne more names that any other group of stars. Babylonian astronomers depicted them as Gilgamesh, the mythical king of ancient Mesopotamia and hero of one of the earliest known works of literature. Phoenicians along the Mediterranean coast associated the stars with their sea-god Melkarth, which remains the name of the mu (μ) star in the constellation. In most accounts, the hero has tired of his grand labors, accounting for his faint appearance in the sky.

MAIN STARS

β | KORNEPHOROS
COLOR: Yellow
MAGNITUDE: 2.8
DISTANCE (LY): 148

ζ | ZETA HERCULIS
COLOR: Yellow
MAGNITUDE: 2.8
DISTANCE (LY): 35

δ | SARIN
COLOR: White
MAGNITUDE: 3.1
DISTANCE (LY): 79

π | PI HERCULIS
COLOR: Orange
MAGNITUDE: 3.1
DISTANCE (LY): 380

α^1 | RASALGETHI
COLOR: Red
MAGNITUDE: 3.5
DISTANCE (LY): 380

SAGITTARIUS: The Archer

MAKEUP: 22 stars

BEST VIEWED: July/Aug.

LOCATION: Southeast

DEEP-SKY OBJECT: Great Sagittarius Star Cloud

THE ARCHER IS A CENTAUR—half man, half horse—located at the widest band of the Milky Way, and it offers a window to the center of the galaxy. Bordered by zodiac neighbors Scorpius and Capricornus, Sagittarius appears in mid- to late summer in the southern sky as he pursues his prey, Scorpius. The ancient Greeks identified Sagittarius with the centaur Chiron, who also appears in the constellation Centaurus. To find Sagittarius, locate Vega and measure about 50 degrees straight south. Another way to spot Sagittarius is to look for the well-known asterism that falls within it: The eight central stars form the shape of a teapot, and you can think of the Milky Way as steam coming from the teapot's spout.

Bright Stars and Objects

Among the more well-known star forms, Sagittarius is a prime example of how the Bayer system for naming stars does not always reflect the order of magnitude. The cataloging can be off sometimes, because of either mistakes or later changes in constellation borders, and the brighter star turns out to be the beta or gamma star. Sagittarius has plenty of candlepower—eight of its stars are brighter

On one side of the teapot's lid are a group of Messier objects all within reach of binoculars: the Lagoon Nebula (M8); the Omega Nebula (M17), which is likened to a swan; and the Trifid Nebula (M20), a colorful stellar nursery. The Trifid takes its name from the three dark dust lanes that split up the bright nebula and that appear to have distinct blue and red regions in photos. Located some 5,200 light-years away, it remains an easy target in high-power 10 × 50 binoculars and small telescopes.

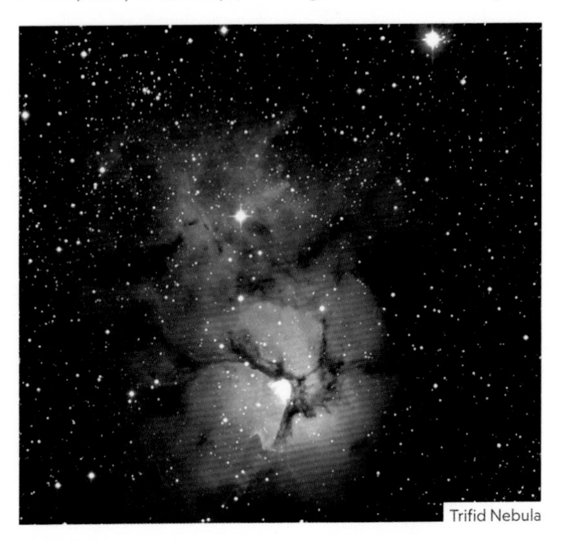

Trifid Nebula

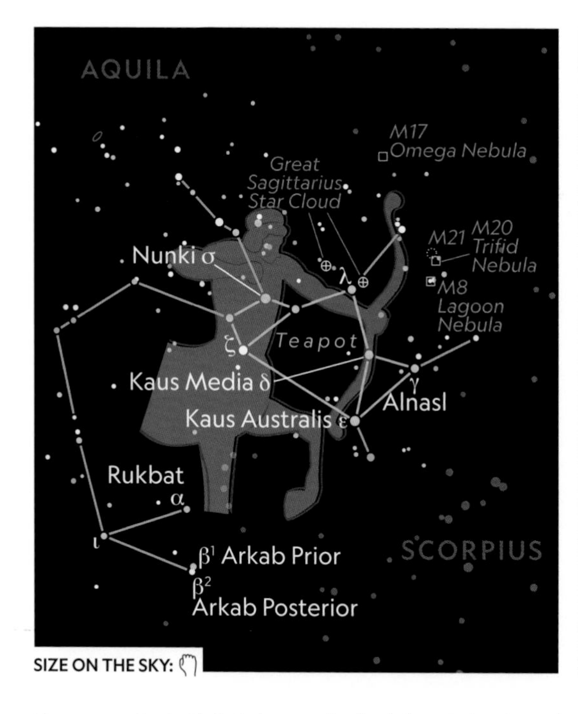

SIZE ON THE SKY:

MAIN STARS

ε | KAUS AUSTRALIS
COLOR: Blue
MAGNITUDE: 1.8
DISTANCE (LY): 145

σ | NUNKI
COLOR: Blue
MAGNITUDE: 2.1
DISTANCE (LY): 228

ζ | ZETA SAGITTARII
COLOR: White
MAGNITUDE: 2.6
DISTANCE (LY): 88

δ | KAUS MEDIA
COLOR: White
MAGNITUDE: 2.7
DISTANCE (LY): 305

λ | LAMBDA SAGITTARII
COLOR: Orange
MAGNITUDE: 2.8
DISTANCE (LY): 78

than magnitude 3. But at magnitude 4 the alpha star, Rukbat, is only the 14th brightest. The brightest star is actually Kaus Australis (Epsilon Sagittarii) at magnitude 1.8. The second brightest star is Sigma Sagittarii, also known as Nunki, prominent enough to be included by the Babylonians on their Tablet of the 30 Stars.

Sagittarius is oriented toward the core of the galaxy, which is some 28,000 light-years away. In this part of the sky, binoculars or a telescope will let you skip among a half dozen or more deep-sky objects. The most striking deep-sky target is the 10,000-light-years-distant Great Sagittarius Star Cloud above the lid of the teapot; it is visible with binoculars, but telescopes will show a countless number of stars. The stars closer to Earth will be the brightest, but this clear patch of sky allows us to see far deeper into the Sagittarius arm of our spiral galaxy. The large hazy patch occupies a space roughly equivalent to nine full moons.

FURTHER

Tea is popular in the Arab world today, but ancient Arab astronomers did not associate Sagittarius's stars with the object that holds it. To them, the bright group of stars that forms the teapot represented ostriches on their way to drink from the Milky Way.

SCORPIUS: The Scorpion

MAKEUP: 17 stars

BEST VIEWED: July/Aug.

LOCATION: Southwest

DEEP-SKY OBJECT: M6, Butterfly cluster

SCORPIUS IS one of the most evocative constellations. Two of the 25 brightest stars in the sky lie within it, surrounded by countless other brilliant stars and deep-sky objects. Northern observers will spot it most easily in summer, toward the south along the Milky Way and between zodiac partners Sagittarius and Libra. Unlike many constellations, Scorpius very much resembles its namesake animal—from its wide, pincer-armed head to a torso represented by brilliant Antares and a twisted tail that ends with the bright star Shaula.

Brightest Stars

For novice stargazers, Scorpius is among the easiest constellations to recognize, because of both its fishhook asterism and magnitude-1 Antares, which sits at its heart. Antares was known to the Romans as Cor Scorpionis, or the "heart of the scorpion." This red supergiant burns 10,000 times brighter than our sun and is about 700 times larger as it nears the end of its stellar life. It has been closely

FURTHER

The name Antares is Greek for "rival of Ares"—testament to the red star's color and prominence in the sky.

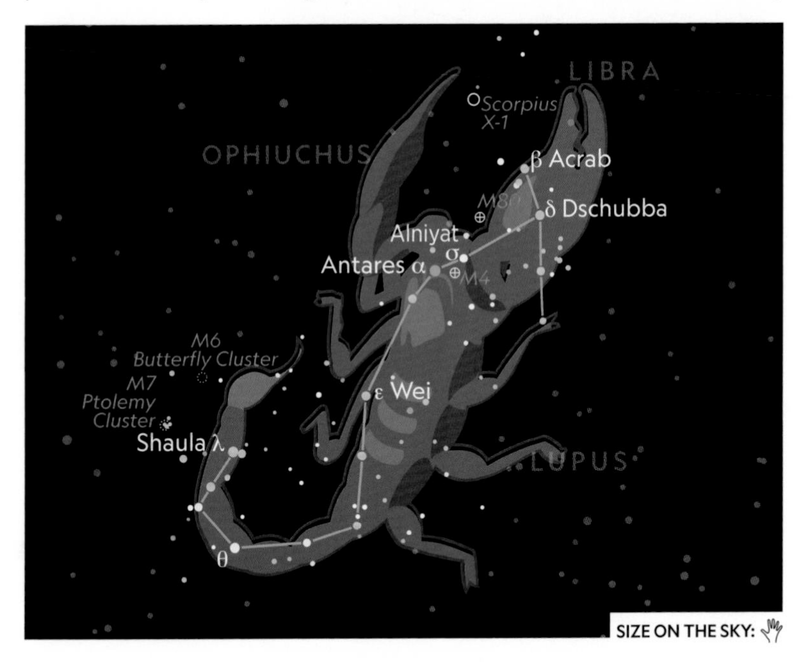

INTERESTING OBJECTS in Scorpius

Scorpius's section of sky is teeming with star clusters. Globular cluster M4 lies just west of Antares. The Butterfly cluster (M6) is a large open cluster that, with the help of binoculars or a telescope, unfolds into the shape of a butterfly. More than 100 stars occupy large open cluster M7, which will come into view with binoculars just east of Shaula. Claudius Ptolemy recognized this group of stars, and thus it is named Ptolemy's cluster. Scorpius X-1 is faint at magnitude 13—it will take some skill to find. This close double star is the brightest x-ray source in the sky. Only one member of the binary can be seen: The partner is a white dwarf, neutron star, or black hole drawing gas away for its neighbor.

Butterfly cluster

watched by humans: The Chinese know it as Ming T'ang, or the "emperor's council hall," while Persians called it Satevis and considered it one of the guardians of the heavens. Shaula, the second brightest star in Scorpius, is a multiple star system with three visible components.

Mythology

This celestial shape has been significant to numerous civilizations. Chinese astronomers saw a mighty dragon. Ancient Greek mythology recognizes Scorpius as the animal that bested the hunter Orion; he was able to inflict the fatality when the hero stepped on its stinger, a wound that glows in the sky as the red star Rigel. The Hunter and the Scorpion are kept separated on opposite sides of the heavens, with Scorpius rising in the east in the spring just as Orion sets in the west.

MAIN STARS

α | ANTARES
COLOR: Red
MAGNITUDE: 1.0
DISTANCE (LY): 600

λ | SHAULA
COLOR: Blue
MAGNITUDE: 1.6
DISTANCE (LY): 570

θ | GIRTAB
COLOR: Yellow
MAGNITUDE: 1.9
DISTANCE (LY): 300

δ | DSCHUBBA
COLOR: Blue-white
MAGNITUDE: 2.2
DISTANCE (LY): 400

ε | WEI
COLOR: Red
MAGNITUDE: 2.3
DISTANCE (LY): 65

DRACO: The Dragon

MAKEUP: 18 stars

BEST VIEWED: May/June

LOCATION: Northeast

DEEP-SKY OBJECT:
NGC 6543, Cat's Eye Nebula

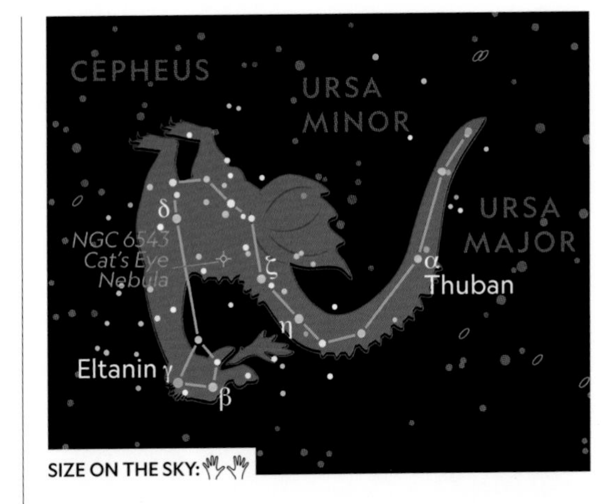

SIZE ON THE SKY:

THIS IS ONE OF THE CLOSEST constellations to the celestial north pole, a giant circumpolar shape that is visible all year in the Northern Hemisphere. Draco's alpha star, Thuban, was once Earth's polestar about 5,000 years ago. Precession has since shifted the celestial position of Earth's axis, and Polaris has taken Thuban's place as the star positioned at the north celestial pole.

Although visible year-round, Draco is easiest to spot in the early summer, with four stars marking out its boxy head. Locate the bright star Vega of the Summer Triangle, then travel slightly north-northwest to find Draco's brightest star Eltanin. The dragon's tail winds between Ursa Minor and Ursa Major, or the two dippers. The Quadrantid meteor shower erupts from the meeting point of Boötes and Draco at the beginning of January. Draconids, which peaks in October, is a minor meteor shower but it erupted into storm levels twice in the 20th century.

Mythology

Draco has represented various scaly beasts to different civilizations. Ancient Greeks considered the constellation to be the dragon Ladon, which was slain by Hercules. Hindu mythology sees an alligator, while the Persians claimed it was a giant serpent.

Combined images from the Hubble Space Telescope and the Chandra X-ray Observatory show that the Cat's Eye Nebula (NGC 6543), a dying, sunlike star in the shape of a bull's eye, is emitting extremely hot gases. Small backyard telescopes show the nebula as a tiny bluish disk 3,600 light-years away.

OPHIUCHUS: The Serpent Bearer

MAKEUP: 14 stars

BEST VIEWED: June/July

LOCATION: Center of chart

DEEP-SKY OBJECT: RS Ophiuchi

OPHIUCHUS AND SERPENS were once one constellation, but they have since been split. The mythologies of the two constellations are as intertwined as their stars. Ophiuchus is easy to identify, lying just west of the Milky Way above Scorpius and below Hercules. While he does sit on the ecliptic, Ophiuchus is not in the Zodiac. There is a series of globular clusters spread out within it—M9, M10, M12, M14, M19, M62, and M107—that are all visible through binoculars. RS Ophiuchi ranges in magnitude from 11.8 to 4.3 and is a cataclysmic variable, a binary system of a white dwarf orbiting with a normal star that is constantly transferring mass to its more powerful companion. Barnard's Star, one of the nearest stars to Earth, is a magnitude-9.5 star that is visible with binoculars. Moving across the sky at 103 miles a second (166 km/s), it has the greatest proper motion of any known star.

Mythology

Ophiuchus is thought to honor Asclepius, the god of medicine. The snake wrapped around him, Serpens, taught the healer about the properties of plants, and he soon became skillful enough to raise the dead. His herbal remedies raised Orion after a fatal wound from Scorpius. Hades, god of the dead, became concerned and convinced Zeus to kill Asclepius. Now the god and his serpent watch over us from the sky.

Globular cluster M14

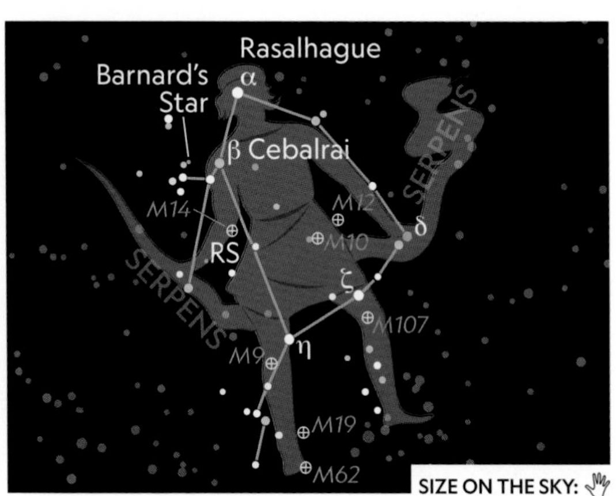

SIZE ON THE SKY:

SERPENS: The Serpent

MAKEUP: 8 stars (Caput) & 6 stars (Cauda)

BEST VIEWED: June/July

LOCATION: Southern half of chart

DEEP-SKY OBJECT: M16, Eagle Nebula

THIS CONSTELLATION winds behind Ophiuchus, the Serpent Bearer, toward the south. The snake is the only divided constellation. The front group of stars including the triangular head is the more prominent section called Serpens Caput, while on the other side is Serpens Cauda, which represents the serpent's tail. Serpens Cauda arcs in the space between bright stars Altair and Antares.

Stars & Objects

Barely detectable with the naked eye, globular cluster M5 is quite beautiful through a small telescope. You can hunt it down about one-third of the way between Arcturus and Antares. Sitting 25,000 light-years distant, this city of a half million stars is one of the largest clusters, spanning 165 light-years across. Astronomers estimate its age at 13 billion years, making it one of the oldest globulars of those scattered in a halo around the Milky Way.

A hidden, not-to-be-missed gem in Serpens is Graff's cluster (IC 4756). Taking up more space in the sky than the

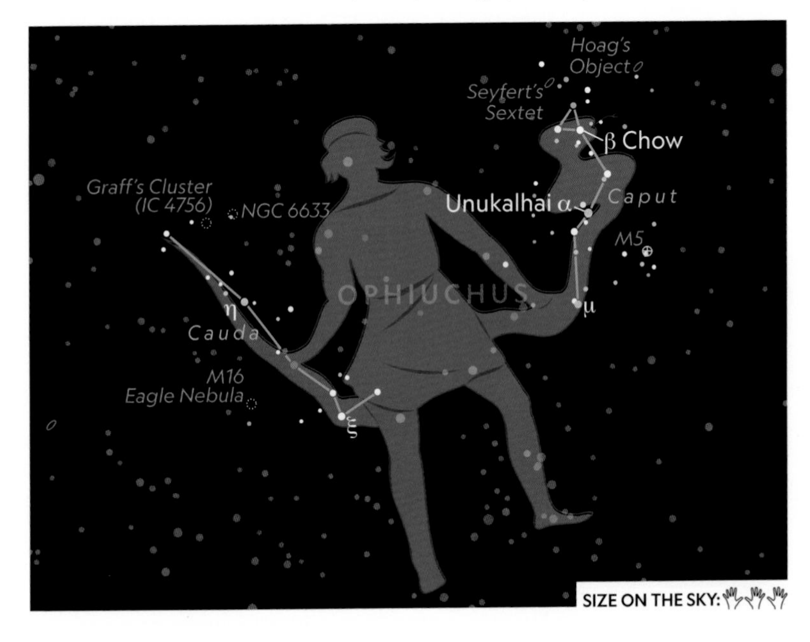

INTERESTING OBJECTS in Serpens

The Eagle Nebula (M16) in Serpens Cauda, 7,000 light-years away, is a combination nebula and star cluster. Hot, energetic young stars no more than six million years old are embedded inside the nebula and light it up. Two can be spotted with an 8-inch (200 mm) telescope or larger. However, even binoculars will give glimpses of the 20 brightest stars that make up the cluster and the faint, wispy hints of the gas cloud that surrounds them. Dark pillars of dense material rise in the center of the nebula; these can be seen with a 12-inch (300 mm) telescope.

Eagle Nebula

moon's disk, this magnitude-4 cluster sits 1,300 light-years away and is barely visible with the naked eye. However, binoculars will reveal a bewildering number of stars across the entire field of view. Just next door, you can spot the equally pleasing open star cluster (NGC 6633). Open clusters like these are valuable for studying the life cycles of stars because their members are all the same distance and age: around 600 years.

Another notable object is Seyfert's Sextet, which is relatively faint though it contains several galaxies. The galaxies are packed so closely together that their gravitational pull has started to rip stars away from each other and in billions of years they may all converge.

Mythology

The snake in this constellation is supposed to have taught medicine to the Greek god Asclepius, which he used to restore the dead to life. The other gods disapproved, so they killed him and sent him and the serpent to the sky.

MAIN STARS

α | UNUKALHAI
COLOR: Orange
MAGNITUDE: 2.7
DISTANCE (LY): 73

η | TANG
COLOR: White
MAGNITUDE: 3.2
DISTANCE (LY): 62

μ | LEIOLEPIS
COLOR: White
MAGNITUDE: 3.5
DISTANCE (LY): 156

ξ | NEHUSHTAN
COLOR: Yellow
MAGNITUDE: 3.5
DISTANCE (LY): 105

β | CHOW
COLOR: White
MAGNITUDE: 3.7
DISTANCE (LY): 153

SCUTUM: The Shield

MAKEUP: 4 stars

BEST VIEWED: July/Aug.

LOCATION: Southeast

DEEP-SKY OBJECT:
M11, Wild Duck cluster

SCUTUM IS another one of the constellations named by Johannes Hevelius in the late 17th century. It is a modern constellation, but its name and shape invoke historical events—perhaps the only constellation with contemporary (at least to Hevelius) political overtones. The original name for this constellation was Scutum Sobiescanum, which translates as "shield of Sobieski." John III Sobieski was the king of Poland and commander of the forces that defeated the Ottoman Empire in the critical 1683 Battle of Vienna. King John was called upon to lead 70,000 men against an army twice that size. It may seem odd for Hevelius to have named a constellation after a man who was still alive, but Sobieski—along with repulsing what was considered a major threat to Europe—was also one of the astronomer's financial backers.

Bright Stars and Objects

The Shield can be found on the Milky Way between Sagittarius, Aquila, and the lower half of Serpens. Scutum is

MAIN STARS

α | IOANNINA
COLOR: Orange
MAGNITUDE: 3.8
DISTANCE (LY): 199

β | BETA SCUTI
COLOR: Orange
MAGNITUDE: 4.2
DISTANCE (LY): 920

ζ | ZETA SCUTI
COLOR: Yellow
MAGNITUDE: 4.7
DISTANCE (LY): 207

γ | GAMMA SCUTI
COLOR: White
MAGNITUDE: 4.7
DISTANCE (LY): 319

δ | DELTA SCUTI
COLOR: Yellow
MAGNITUDE: 4.7
DISTANCE (LY): 202

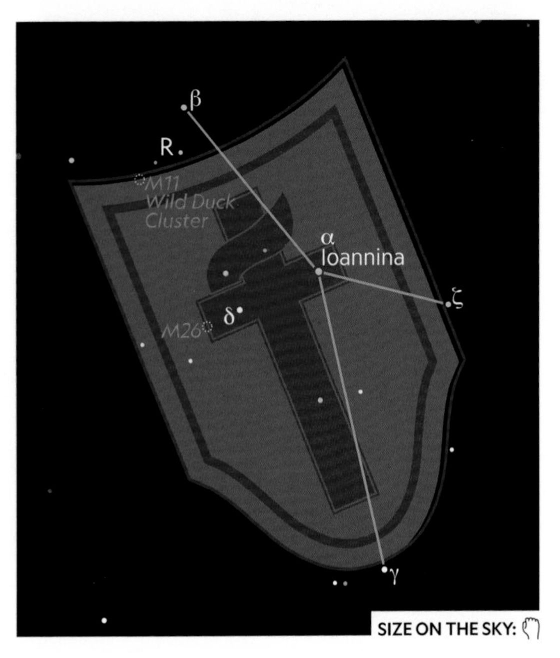

SIZE ON THE SKY:

Wild Duck cluster

not large, nor does it have any very bright stars, but its distinguished company makes it a good viewing target in the Northern Hemisphere in the summer months.

Scutum is home to a pulsating variable star called R Scuti. While this 3,600-light-years-distant yellow supergiant spends much of its time shining at around magnitude 5, it can plunge down to magnitude 8 every four to five months. For backyard sky-watchers, R Scuti is a great variable to follow—it's always within reach of binoculars but can at times be bright enough to spot with the unaided eye.

Just slightly southeast of Delta Scuti (δ) sits the open cluster M26. Estimated at 5,000 light-years from Earth, it's a fine magnitude-8 target for binoculars; however, it may be a bit of a challenge to identify among the surrounding star field because the cluster's members are so scattered.

Astronomers are particularly interested in the shell of Supernova G21.5-0.9 located about 20,000 light-years away in Scutum. We have known about the supernova for a number of years, but between 1999 and 2004 the Chandra X-ray Observatory was able to highlight the outer shell of ejected x-ray material.

INTERESTING OBJECTS in Scutum

Though small, Scutum's position in the Milky Way makes it home to one of the galaxy's highlights: the Wild Duck cluster (M11). So named because of how much it looks like a thick flock of waterfowl, the Wild Duck cluster southwest of the constellation's northernmost star (Beta Scuti) is magnificent through binoculars. Just a 4-inch (100 mm) telescope will start to show the 6,200-light-years-distant cluster's compact V-shape, but it becomes more impressive through an 8-inch (200 mm), with hundreds of glittering stars spread out across 25 light-years.

VULPECULA: The Fox

MAKEUP: 3 stars

BEST VIEWED: Aug./Sept.

LOCATION: Southeast

DEEP-SKY OBJECT:
M27, Dumbbell Nebula

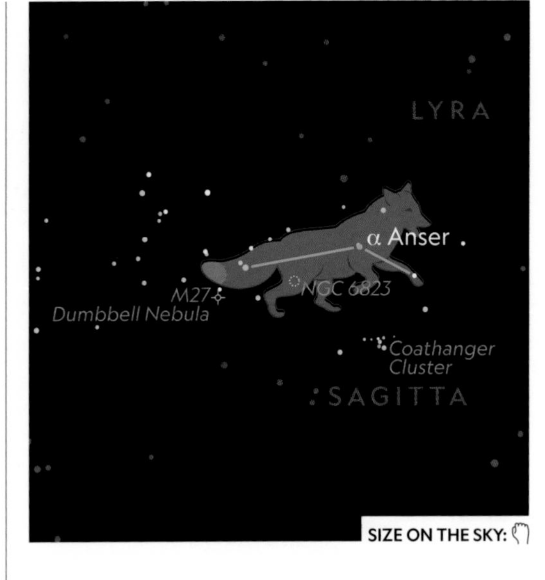

THE FOX IS FOUND high in the late summer sky, roughly in between the bright stars Altair in Aquila and Vega in Lyra, inside the band of the Milky Way. It is a challenging constellation to pick out because it is far dimmer than its surroundings.

Stars & Objects

Vulpecula's stars may be faint, but near the Fox is a fine example of an elegant, even startling, find: the Coathanger (or Brocchi's) cluster, an asterism on the border of Sagitta. On a dark night, it is visible to the naked eye, but binoculars and small telescopes resolve it into individual stars—a straight line of six with a "hook" of four in the middle. First recorded by astronomer Al-Sufi back in A.D. 964, first-time viewers are invariably jolted to see such an artificial-looking arrangement of stars. Astronomers today believe the individual stars that make up this asterism are actually at vastly different distances and so have no relation to each other. Vulpecula does host a real open cluster: NGC 6823, just southwest of Anser, the brightest star in the constellation.

INTERESTING OBJECTS in Vulpecula

In 1764, Charles Messier added the first record of a planetary nebula, M27, to his list of objects that are not comets, but it was John Herschel who nicknamed it the Dumbbell Nebula. M27 is known as a gaseous emission nebula and takes its name from its double- lobed appearance. A good telescope will reveal the shape behind the object's nickname, but it can be spotted as a faint, fuzzy patch through binoculars just north of the brightest star in Sagitta.

SAGITTA: The Arrow

MAKEUP: 6 stars

BEST VIEWED: Aug./Sept.

LOCATION: Southeast

DEEP-SKY OBJECT:
M71, globular cluster

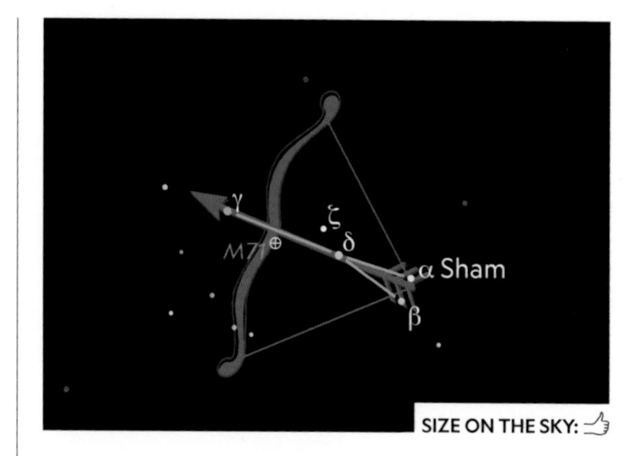

SIZE ON THE SKY:

THIS SMALL, BUT DISTINCT CONSTELLATION is inside the Milky Way on a direct line traced between bright stars Altair of Aquila and Deneb of Cygnus. Sagitta is one of the smallest constellations and one of Ptolemy's original 48. Located very close to the celestial equator, the Arrow can be seen from almost any location on Earth. The entirety of this small constellation lies within the Summer Triangle asterism of Altair, Deneb, and Vega. Sky-watchers will find it high in the sky during the summer months, which is when it is best seen in mid-northern latitudes. A deep-sky object of interest is M71, a globular cluster that's easy to see through binoculars in the arrow's shaft. This bright cluster is located about 12,000 light-years from Earth and measures some 27 light-years across.

Globular cluster M71

Mythology

Although Sagitta is small and fairly dim, many cultures—including the Greeks, Romans, Persians, and Hebrews—designated this group of stars as an arrow. In Greco-Roman myth, Sagitta is said to have been several different arrows: the one used by Cupid in the art of conjuring love, by Apollo in slaying the Cyclops, and by Hercules in dispatching the Stymphalian Birds.

LIBRA: The Scales

MAKEUP: 8 stars

BEST VIEWED: June/July

LOCATION: Southwest

DEEP-SKY OBJECT:
Delta Librae, variable star

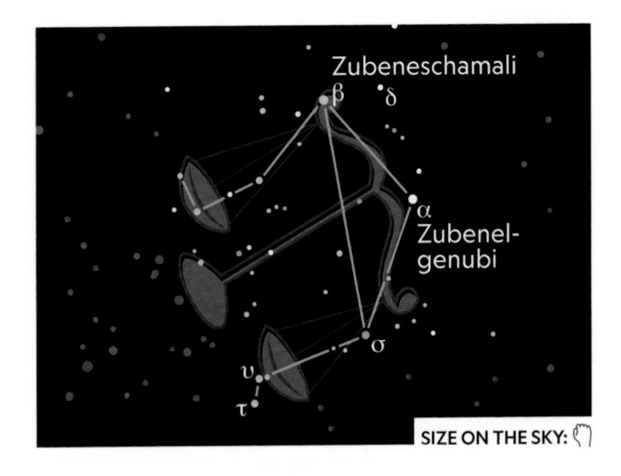

LIBRA'S THREE BRIGHTEST STARS form a triangle at the top of the scales, and the two pans of the balance fall westward toward the bright star Antares in the constellation Scorpius. In the northern summer sky it will be just above the Milky Way, lying close to the southern horizon between its neighbors on the zodiac, Virgo and Scorpius.

Delta Librae (δ), found just west of the northern end of the beam, is of viewing interest. This star is an eclipsing variable whose 2.3-day cycle ranges from magnitude 4.9 to 5.9, perfectly visible to the naked eye on a clear night. Astronomers claim stars cannot be green colored, but Zubenelgenubi, or Alpha Librae—the second brightest star in Libra—seems to defy them. It's a bit of a viewing mystery, as many observers over the centuries have pointed out that it is the only one in the sky to have a distinctive green hue. What color do you see?

Mythology

Arab astronomers named the two brightest stars in this constellation, Zubenelgenubi and Zubeneschamali, to mark the northern and southern claws of nearby Scorpius. Apparently, it was not until 1,000 years later that the Romans placed the two stars in Libra, the constellation that marked the position of the sun during the autumn equinox when day and night are in equal balance.

FURTHER

Libra is one of the 12 constellations of the zodiac, and it's the only one that doesn't depict a human or an animal. According to myth, the scales are those held by the Greek goddess of justice.

CAMELOPARDALIS: The Giraffe

MAKEUP: 5 stars

BEST VIEWED: Year-round

LOCATION: North

DEEP-SKY OBJECT: NGC 1502, star cluster

THIS CONSTELLATION WAS INTRODUCED in the 1600s to fill the open space between the two bears (Ursa Major and Minor) and Perseus. To the ancient Greeks it was a "camel-leopard," a beast with the head of a camel and the spots of a leopard. Star cluster (NGC 1502) makes for a good telescope target. Stretching toward the cluster is a string of stars known as Kemble's Cascade, with the cluster containing two easily split double stars—one pair at magnitude 5, the other at 9.

An irregular variable star, VZ Camelopardalis, lies northwest near Polaris. Its variation is slight and may be challenging to observe. The magnitude of VZ Camelopardalis, a cataclysmic variable star, ranges from 9.6 to 13 and is visible every night of the year. Another object is NGC 2403, a spiral galaxy about eight million light-years from Earth. A supernova was discovered here in 2004.

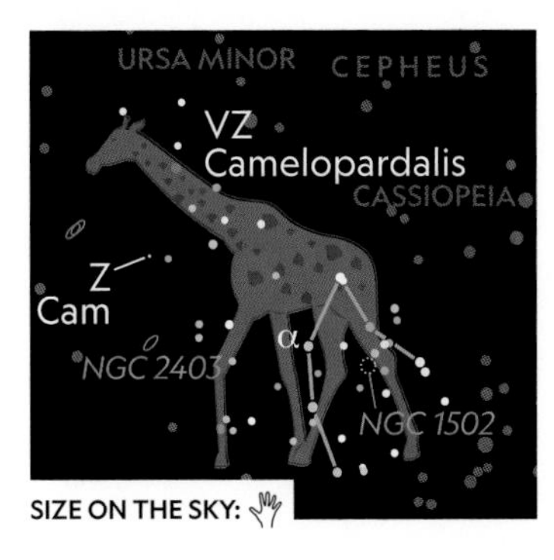

SIZE ON THE SKY:

CORONA AUSTRALIS: The Southern Crown

MAKEUP: 7 stars

BEST VIEWED: July/Aug.

LOCATION: Southeast

DEEP-SKY OBJECT: NGC 6541, star cluster

ONE OF PTOLEMY'S ORIGINAL 48 constellations, Corona Australis rises only a few degrees above the horizon from mid-northern latitudes, just west of the bright star Shaula at the tail of Scorpius. Even then the Southern Crown is still a rather faint constellation, with no star brighter than magnitude 4. Though small, the Southern Crown hosts an active star-forming region: The Coronet cluster and the Corona Australis Nebula are invisible to all but the most advanced deep-sky imaging technology, but photographs show around 30 "newborn" stars. Near the constellation's southern tip, NGC 6541 is a magnitude-6 globular cluster and visible through a small telescope.

For a small constellation, the Southern Crown has a surprising number of associated myths. Most refer to a crown of laurel or fig leaves, which some believe belonged to Chiron, the Centaur.

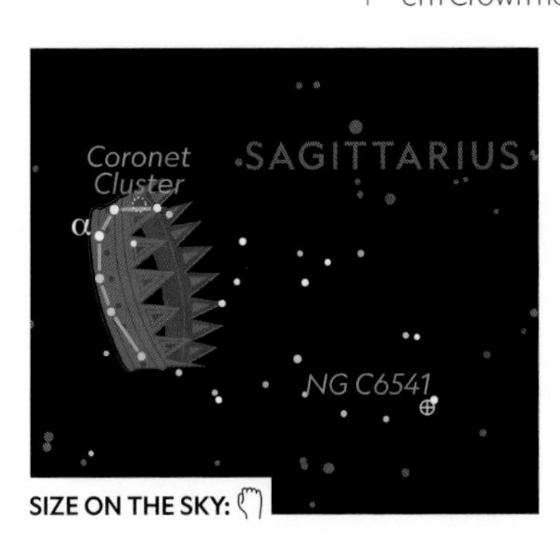

SIZE ON THE SKY:

AUTUMN

THE GREAT SQUARE of Pegasus, an asterism at the center of the constellation of the same name, rides high in the southern sky in late autumn evenings. Bright star Alpheratz is shared by Andromeda and the equine asterism: It leads east, and then north to Perseus and the bright orange star Capella. Below Pegasus, in the southeast, is the

DATE	TIME
9/21	11 p.m.
10/21	10 p.m.
11/1	9 p.m.

winding figure of Pisces and the sprawling, monster whale constellation of Cetus. Look for four major meteor showers in the fall: the Orionids in October, the Taurids and Leonids in November, and the Geminids in December. The season also provides an opportunity to view galaxies, like the Andromeda galaxy (M31), that are visible to the naked eye.

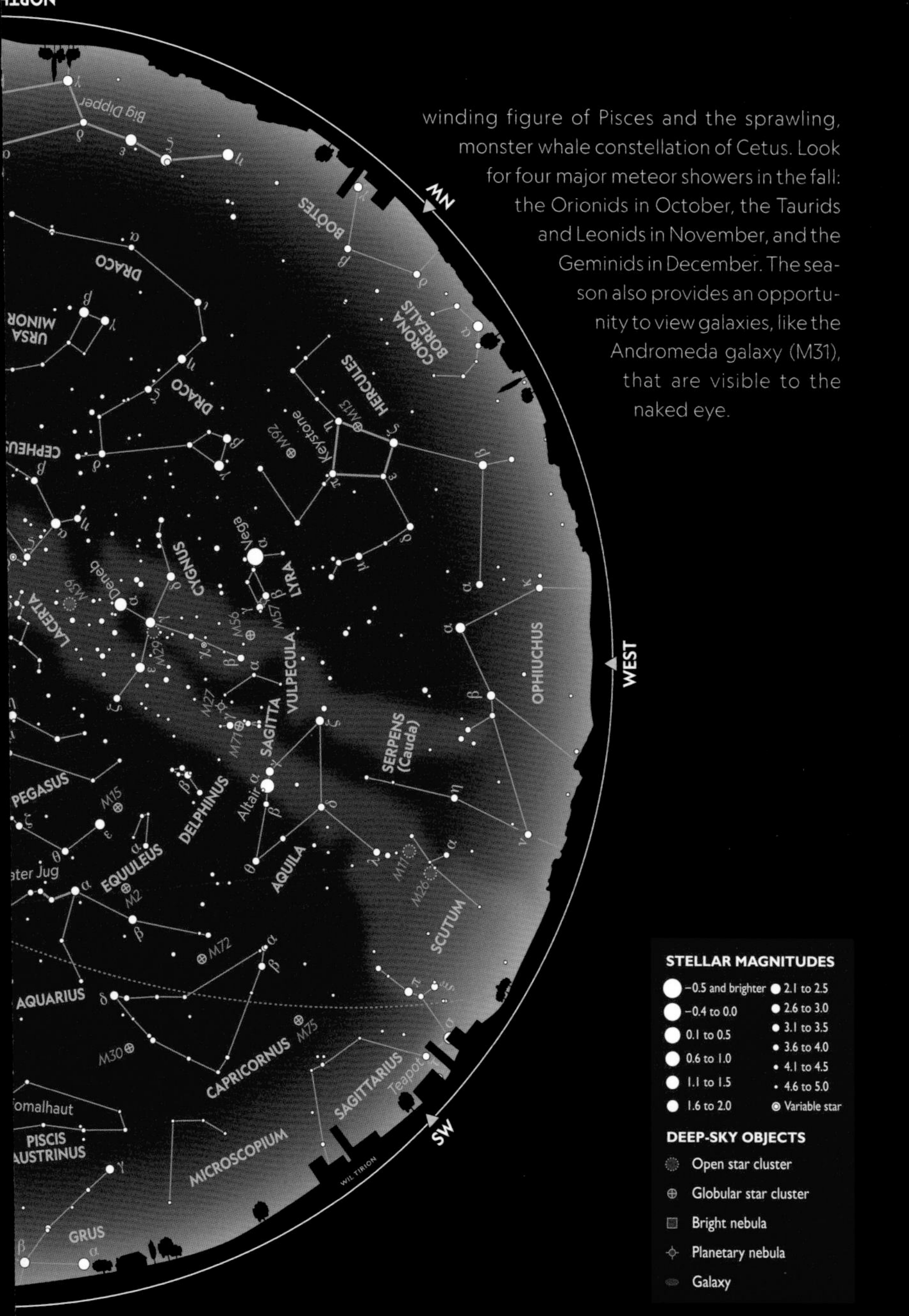

AUTUMN STAR HOPPING

WHILE THE autumn sky lacks super-bright stars, the four stars that form the recognizable square in Pegasus make a stellar signpost that helps sky-watchers hop to several other seasonal constellations, including Andromeda. Extending one side of the Great Square straight south will point the way to Diphda, the brightest star of Cetus (the Sea Monster). Do the same with the other side of the Square and you arrive at Fomalhaut, the lead star in Pisces Austrinus.

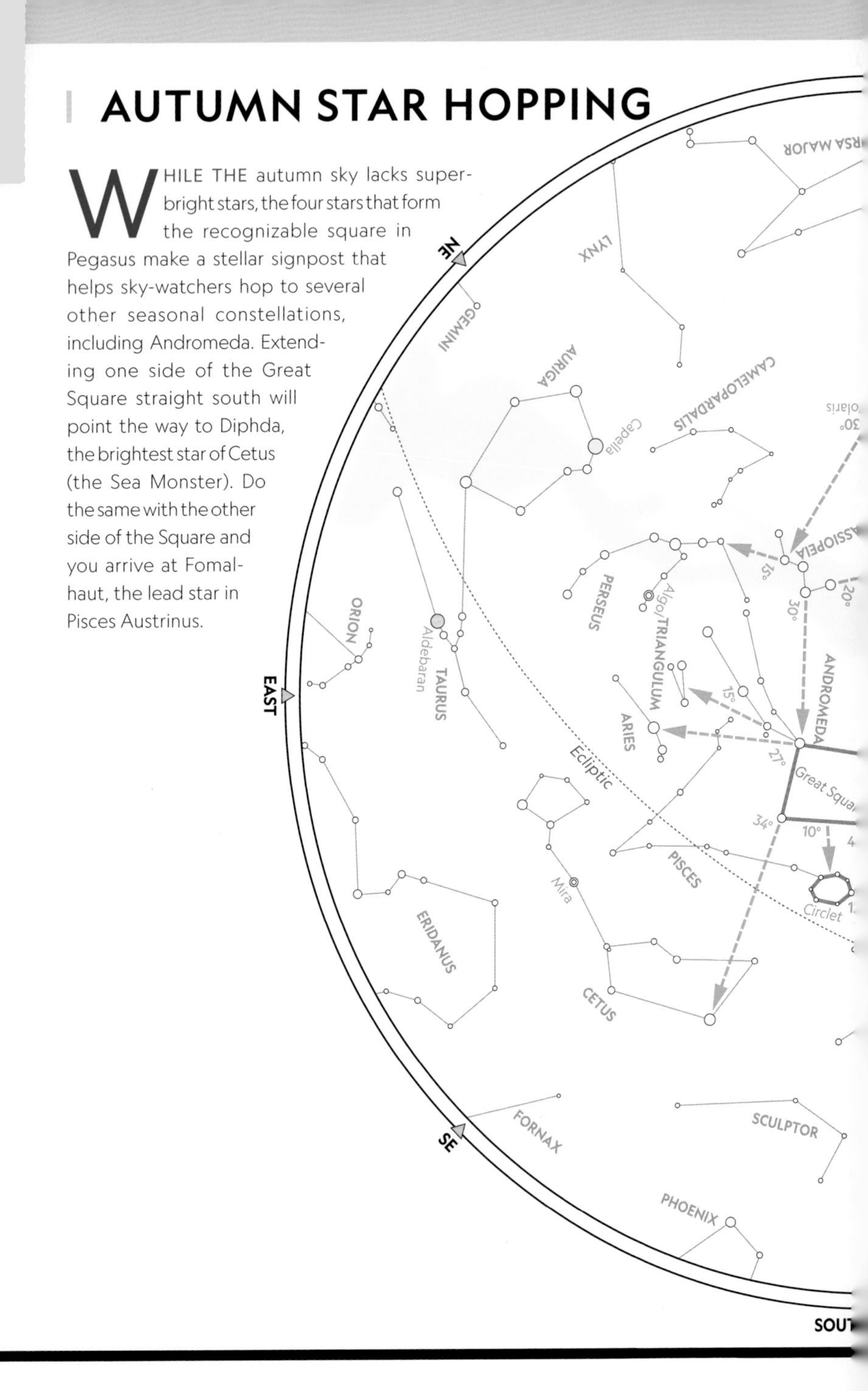

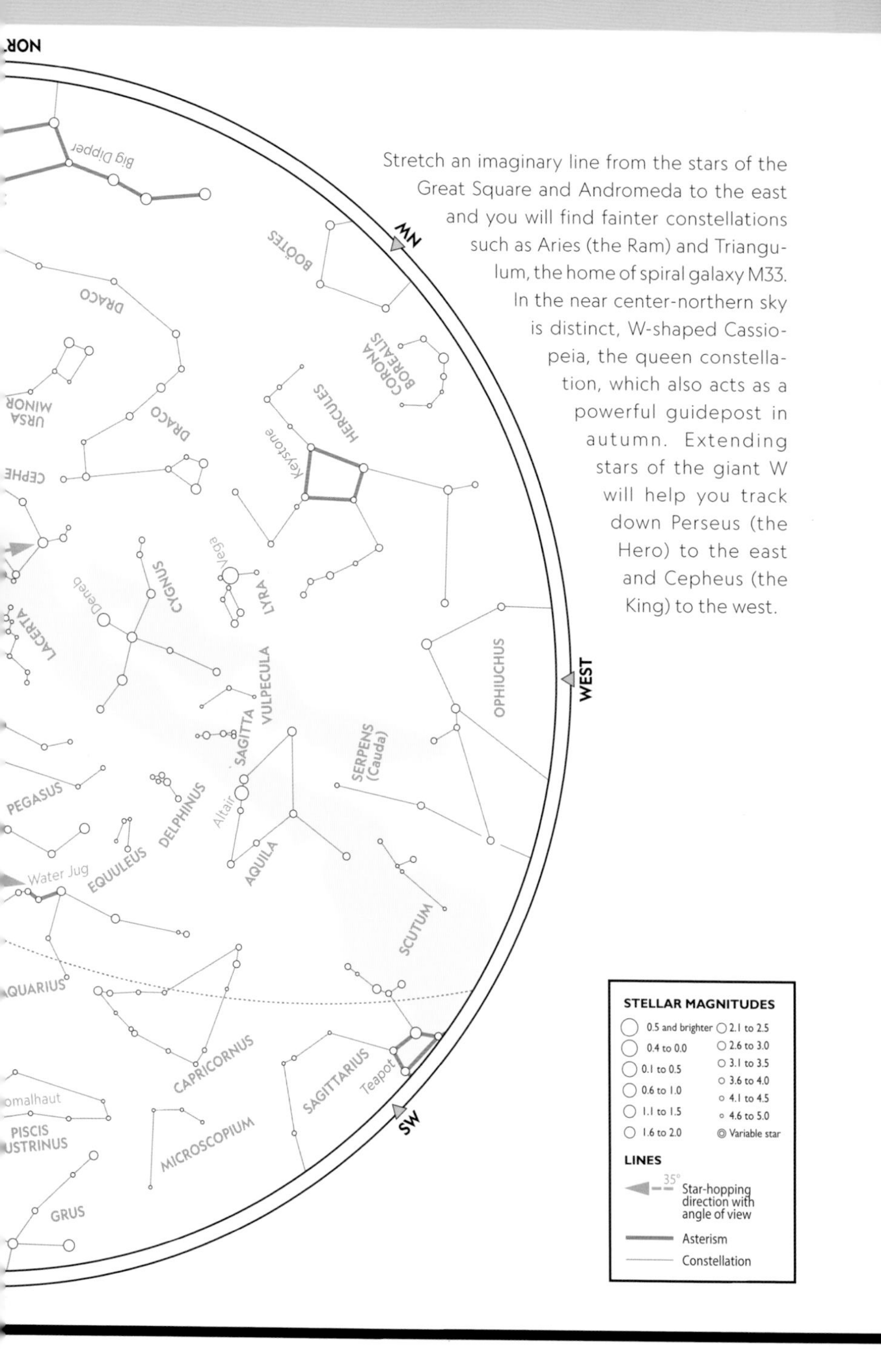

Stretch an imaginary line from the stars of the Great Square and Andromeda to the east and you will find fainter constellations such as Aries (the Ram) and Triangulum, the home of spiral galaxy M33. In the near center-northern sky is distinct, W-shaped Cassiopeia, the queen constellation, which also acts as a powerful guidepost in autumn. Extending stars of the giant W will help you track down Perseus (the Hero) to the east and Cepheus (the King) to the west.

CASSIOPEIA: The Queen

MAKEUP: 5 stars

BEST VIEWED: Oct./Nov.

LOCATION: Northeast

DEEP-SKY OBJECT: M52, open star cluster

CASSIOPEIA IS VISIBLE all year in the Northern Hemisphere because of her proximity to the north celestial pole. She sits on her throne facing-away from Polaris, within a portion of the Milky Way. Beneath her is Andromeda, and above her is Ursa Minor (the Little Bear). Cassiopeia's obvious W shape (or M, depending on the season) makes this constellation one of the easiest to identify. The best time to observe Cassiopeia is in the fall, when she reaches her highest point.

One of several star clusters in the constellation, 5,000-light-years-distant open cluster M52 is the best one for a novice stargazer to view, as it is easily visible through a telescope. Locate it by following the path of the southernmost leg of the W. The cluster contains about 100 stars,

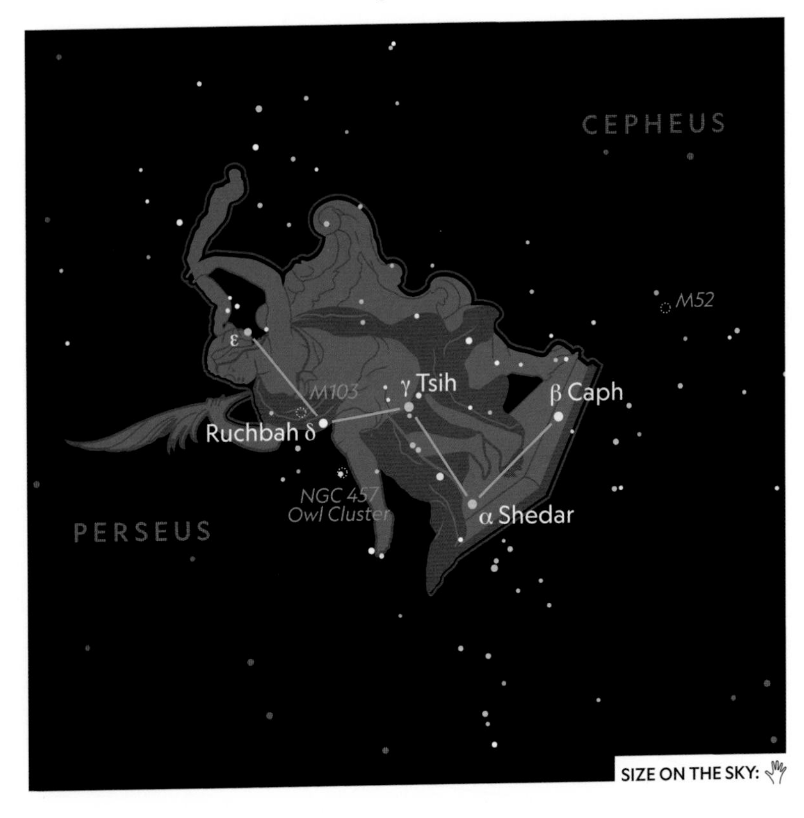

INTERESTING OBJECTS
in Cassiopeia

The most visually impressive star cluster in Cassiopeia is NGC 457—not because of its brilliance, but rather its distinctive shape. This 7,900-light-years-distant cluster is also called the Owl cluster and the E.T. cluster. With two bright stars that look like eyes and two streams of stars angling out from its core, some people see owl wings, while others see the outstretched arms of the famous movie alien. Only the brightest stars are visible through binoculars, but small telescopes will reveal fainter ones too.

Owl cluster (NGC 457)

making it one of the richest in the northern sky. Galaxy M103, located one degree from the star Ruchbah, is also visible with binoculars.

Mythology

Cassiopeia is one of the key figures in the Greek myth of Perseus and Andromeda. In the story, Cassiopeia begins by boasting that her daughter's beauty is greater than that of the Nereids, the daughters of the sea god Nereus. For offending the gods of the sea, Cassiopeia is forced to sacrifice her daughter's life or else see the entire kingdom destroyed by a sea monster. Andromeda escapes her fate when Zeus's son, Perseus, rescues her, but Cassiopeia does not avoid punishment. Upon her death, the queen was banished to the sky where she suffers chained to her throne, forced to hang upside down for half the year. Nearby, other characters from the famous tale accompany the boastful Ethiopian queen, including her husband Cepheus.

MAIN STARS

α | SHEDAR
COLOR: Orange
MAGNITUDE: 2.2
DISTANCE (LY): 230

β | CAPH
COLOR: Yellow
MAGNITUDE: 2.3
DISTANCE (LY): 54

γ | TSIH
COLOR: Blue
MAGNITUDE: 2.5
DISTANCE (LY): 610

δ | RUCHBAH
COLOR: White
MAGNITUDE: 2.7
DISTANCE (LY): 99

PERSEUS: The Hero

MAKEUP: 14 stars

BEST VIEWED: Nov./Dec.

LOCATION: Northeast

DEEP-SKY OBJECT: NGC 869 & 884, double cluster

VISIBLE FOR MUCH OF THE YEAR in the northern sky, Perseus straddles the Milky Way between Cassiopeia and Andromeda, characters from the same Greek myth in which he takes center stage as the story's hero. This constellation is relatively easy to pick out for a beginning stargazer. Although visible for much of the year, the best times to view Perseus are in the late fall months.

Bright Stars and Objects

With magnitude-1.8 Mirfak at its center, and five other stars of magnitude 3 or greater, Perseus contains bright and notable signposts. Among them is Algol (Beta Persei), a well-known eclipsing variable that dims by a full magnitude for about 10 hours every three days. Perseus contains a fascinating deep-sky object, the double cluster NGC 869 and 884. It presents the illusion of a large, connected field but is in fact two separate, unassociated groups of stars.

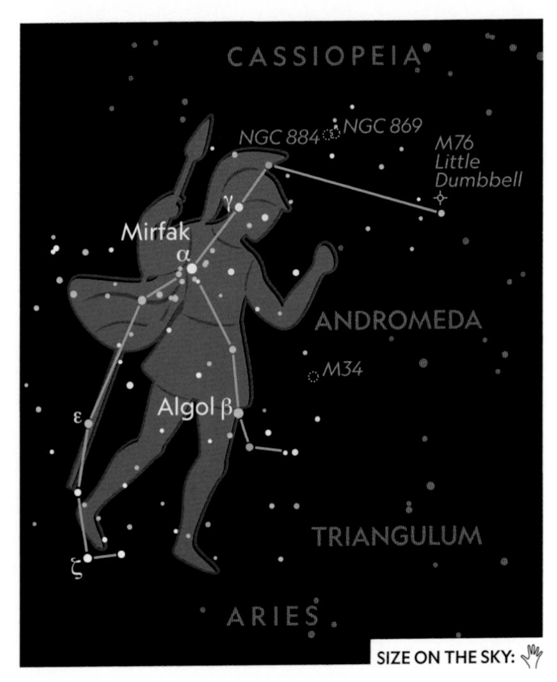

FURTHER

In the story of Perseus, the variable star Algol represents the eye of Medusa. It translates as "the Demon's head" in Arabic and as "Satan's head" in Hebrew. Looks like the ancients suspected that something about this star was foreboding.

Little Dumbbell Nebula (M76)

INTERESTING OBJECTS in Perseus

Located 2,500 light-years from Earth, the Little Dumbbell Nebula (M76) takes its name from its bigger, brighter planetary nebula cousin. This magnitude-10 fuzzy spot is fairly easy to hunt down with small telescopes, since it lies near the border with Andromeda and just south of Cassiopeia.

They are best studied with binoculars or a lower-magnification eyepiece that will provide a wide field of view for this area rich in double and multiple stars.

This constellation is the site of the annual Perseid meteor shower, visible every summer starting in late July, that has been observed for almost 2,000 years. The meteors are debris shed by comet Swift-Tuttle, first spotted in 1862 and with an orbit of 120 years—it last passed by Earth in 1992.

Mythology

Perseus, the half-mortal son of Zeus, is one of the more prominent Greco-Roman heroes. A favorite of Athena, he received many gifts from her, including a polished shield. In going into battle with the Gorgon Medusa—who, with one look, could turn any human or animal into stone—Perseus used the shield as a mirror to avoid her direct gaze. When he lopped off her head, the winged horse Pegasus was born of her blood and Perseus took it for his own. In his second great feat, Perseus used Medusa's head as a weapon to help him rescue Andromeda from the sea monster Cetus: Even severed from her body, the Gorgon's stare turned Cetus to stone.

MAIN STARS

α | MIRFAK
COLOR: Yellow-white
MAGNITUDE: 1.8
DISTANCE (LY): 590

β | ALGOL
COLOR: Blue-white
MAGNITUDE: 2.1
DISTANCE (LY): 93

ζ | ATIK
COLOR: Blue-white
MAGNITUDE: 2.8
DISTANCE (LY): 750

ε | ADID AUSTRALIS
COLOR: Blue
MAGNITUDE: 2.9
DISTANCE (LY): 538

γ | ALPHECHER
COLOR: Yellow
MAGNITUDE: 2.9
DISTANCE (LY): 225

PEGASUS: The Winged Horse

MAKEUP: 15 stars

BEST VIEWED: Sept./Oct.

LOCATION: Center of chart

DEEP-SKY OBJECT:
M15, globular cluster

PEGASUS CERTAINLY EARNED its celestial immortality. The winged horse has a large presence in Greek mythology and despite its relative faintness occupies a large piece of celestial real estate.

Stars & Objects

The mythical winged steed is found in the southern sky right by Andromeda, and the star Alpheratz (Alpha Andromedae) is actually in the princess constellation. This shared star joins with the horse's three brightest stars to form the Great Square of Pegasus, an especially useful asterism that serves as a signpost for star hopping to constellations in the regions close by.

The M15 cluster is one of the more prominent in the Northern Hemisphere, located on the far eastern border of the constellation near the star Enif, which marks Pegasus's nose—the 32,000 light-years-distant cluster represents a pesky fly. This compact ball of stars is visible through binoculars, but you need a telescope to see the main attraction—some 100,000 stars in a tight, sparkling ball.

FURTHER

In a sense, Pegasus is the offspring of the monster Medusa. When Perseus slew her, Pegasus was born of her blood.

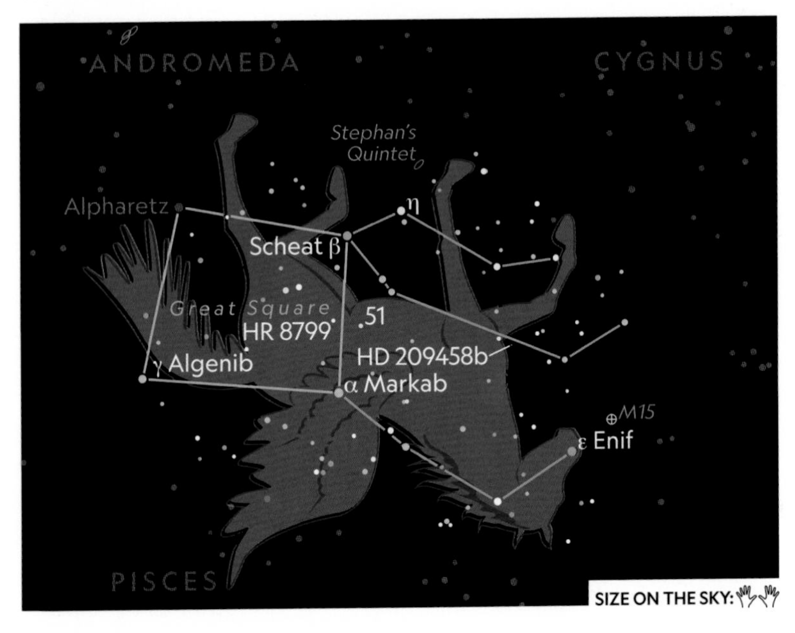

INTERESTING OBJECTS in Pegasus

Pegasus has been a repeated source of excitement in the search for exoplanets—worlds circling stars other than the sun—and a key target in the hunt for extraterrestrial life. In 1995 scientists detected the first exoplanet found around a stable star, 51 Pegasi: a "Hot Jupiter" gas giant orbiting close to its sunlike star. Eleven other stars in Pegasus are now known to have planets. These include HD 209458b, which provided the first evidence of water vapor in an exoplanet atmosphere, and the world's first directly imaged exoplanets captured orbiting HR 8799.

Globular cluster M15

Of general interest is Stephan's Quintet, a group of five galaxies visually close to one another near Pegasus's northern border. Four are in the process of colliding, while the fifth may be a foreground object closer to the Milky Way.

Mythology

Pegasus is a well-known mythological creature. A myth shared with the surrounding constellations Andromeda, Perseus, Cetus, Cepheus, and Cassiopeia refers to the story of Perseus and Andromeda, in which the winged horse serves as Perseus's steed. In another tale, the Greek hero Bellerophon was given a special bridle by Athena in order to tame the flying horse and the pair went on to fight the Chimera monster. Pegasus's final act of service was to carry Zeus's lightning bolts and was rewarded with a place in the sky.

MAIN STARS

ε | ENIF
COLOR: Orange
MAGNITUDE: 2.4
DISTANCE (LY): 690

β | SCHEAT
COLOR: Red
MAGNITUDE: 2.4
DISTANCE (LY): 200

α | MARKAB
COLOR: Blue-white
MAGNITUDE: 2.4
DISTANCE (LY): 140

γ | ALGENIB
COLOR: Blue
MAGNITUDE: 2.8
DISTANCE (LY): 333

η | MATAR
COLOR: Yellow-orange
MAGNITUDE: 2.9
DISTANCE (LY): 167

ANDROMEDA: The Chained Maiden

MAKEUP: 7 stars

BEST VIEWED: Oct./Nov.

LOCATION: Center of chart

DEEP-SKY OBJECT: M31, Andromeda galaxy

IN ANCIENT GREEK MYTH, Princess Andromeda of Ethiopia was chained to a rock as a sacrifice to the gods. In the sky, Andromeda is a faint V-shaped constellation that is most often seen upside down. To locate her on fall evenings, trace a line northeast from the northeast corner of the Great Square of Pegasus.

Stars & Objects

Gamma Andromedae, named Almach, is a stunning golden-yellow-and-blue double star system that is easily split with a small telescope. Only five degrees south of Gamma is the 1,300-light-years-distant open cluster NGC 752, which is best seen through binoculars. Telescopes at least 6 inches (150 mm), when swept 3.5 degrees east of Gamma, will reveal NGC 891—at 33 million light-years away, it's one of the best examples of an edge-on spiral galaxy with a dark dust lane cutting across it.

Most notably, this constellation contains the famous

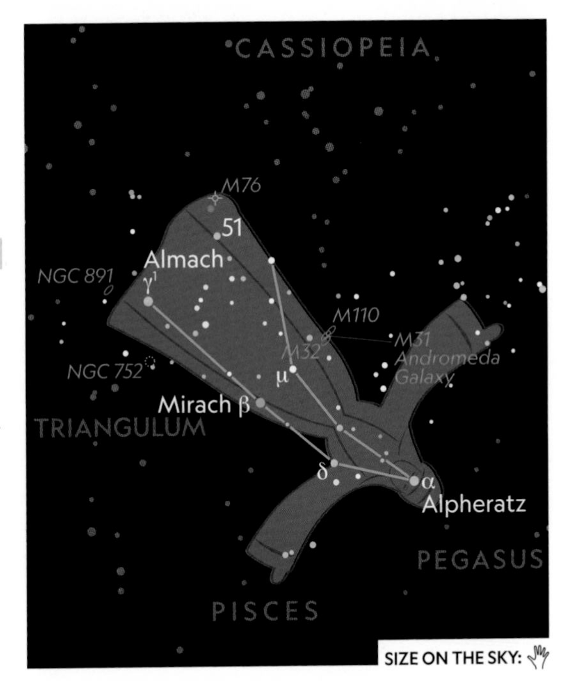

SIZE ON THE SKY:

FURTHER

Finding the Andromeda galaxy is easy: Just star hop from Alpheratz (the corner star of Great Square of Pegasus) two bright stars up to Mirach, and then draw an imaginary line to Mu Andromeda (μ) at a right angle to Mirach. Extend that line out by exactly the same distance as that between those two stars and you'll arrive at Andromeda galaxy.

Galaxy M110, with its bright core, illuminates a patch of sky near other stars in the constellation Andromeda.

Andromeda galaxy (M31). At 2.5 million light-years away, it is one of the most distant objects visible to the unaided eye under dark skies. A giant, spiral-shaped galaxy not unlike the Milky Way, it's estimated to be home to about one trillion stars—more than double the number in our own galaxy. To the naked eye, it looks like an oval smudge just west of Andromeda's right arm. It covers more than three degrees in the sky, which is equal to six moon disks. With binoculars and telescopes you'll be able to see some of the dust lanes beyond its core and the much smaller, fainter elliptical satellite galaxies M32 and M110.

Mythology

Andromeda's mother, Cassiopeia, claimed that her daughter's beauty surpassed that of the daughters of Nereus, a god of the sea and father-in-law of Poseidon. Angered by her boasting, Poseidon sent the monster Cetus to destroy the kingdom of Ethiopia unless Cassiopeia and her husband Cepheus sacrificed their daughter. The princess was rescued by Perseus, who used the head of Medusa to turn Cetus to stone. The whole cast of the tale—her father Cepheus, Cassiopeia, Cetus, Perseus, and the hero's winged horse Pegasus—are all nearby in the sky.

MAIN STARS

α | ALPHERATZ
COLOR: Blue-white
MAGNITUDE: 2.1
DISTANCE (LY): 97

β | MIRACH
COLOR: Orange
MAGNITUDE: 2.1
DISTANCE (LY): 197

γ1 | ALMACH
COLOR: Orange
MAGNITUDE: 2.2
DISTANCE (LY): 390

δ | DELTA ANDROMEDAE
COLOR: Orange
MAGNITUDE: 3.3
DISTANCE (LY): 105

ARIES: The Ram

MAKEUP: 4 stars

BEST VIEWED: Nov./Dec.

LOCATION: Southwest

DEEP-SKY OBJECT:
Gamma Arietis, double star

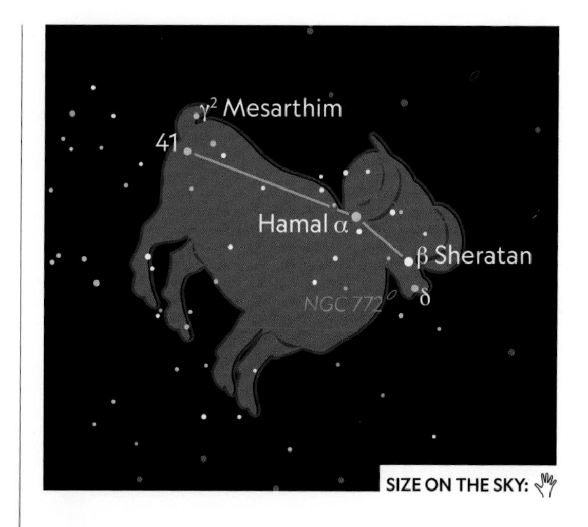

THE ANCIENT ASTRONOMERS who organized the zodiac noted that the sun was "in" Aries at the vernal equinox—meaning it was traveling in the Ram's part of the sky on the day when it passed from the southern to the northern celestial sphere for the year. Precession has since shifted this constellation's position with respect to the sun's equatorial crossing, but by tradition Aries remains where the zodiac begins. On evenings in late fall and early winter, Aries is high in the east between the Great Square of Pegasus and the Pleiades in Taurus. Gamma Arietis, named Mesarthim, is a double star made up of γ^1 and γ^2 with a wide, eight-arc-second separation between the two members.

NGC 772 galaxy has a single prominent spiral arm.

Mythology

Aries was widely recognized as a ram despite being made of only four stars. According to the Greeks, Aries was the source of the Golden Fleece. The ram was sent to rescue a king's children from an abusive stepmother. When the ram returned, the king sacrificed it and left the fleece with a dragon, until it was stolen by Jason.

AURIGA: The Charioteer

MAKEUP: 7 stars

BEST VIEWED: Dec./Jan.

LOCATION: Center of chart

DEEP-SKY OBJECT:
M36 & M37, star clusters

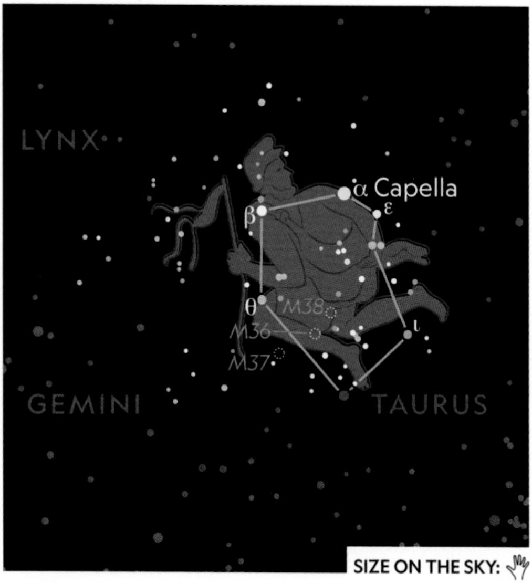

SIZE ON THE SKY:

AURIGA IS AN ELEGANT CONSTELLATION in the heart of the Milky Way. It is easily identified by its alpha star, the brilliant Capella, which is the seventh brightest star in the sky. Auriga is an ancient constellation, one of Ptolemy's original 48. Epsilon Aurigae (ε), the star just southwest of Capella, is an eclipsing binary, veiled every 27 years by an unknown companion. The next eclipse will begin in 2036.

Deep-Sky Objects

Straddling the galactic equator, Auriga provides a window onto several interesting star clusters. One bright highlight is M36, which can be found just five degrees southwest of Theta Aurigae. Another open star cluster, M38, is best seen through a small telescope. Lying 4,800 light-years away, it is the largest of all Auriga's clusters but is lightly populated compared with others in the constellation.

One of the finest open clusters in the northern sky, M37 lies 4,400 light-years away and is about the same size as the moon's disk in the sky. It appears as a fuzzy patch in binoculars, while it is rich with stars in a small telescope.

FURTHER

There are a few stories associated with this constellation. The star Capella represents a mother goat that the charioteer carries on his back along with her three kids (the neighboring stars). Another story has it that the chariot and rider may represent Hephaestus, the crippled blacksmith god who built the vehicle to move about more easily.

AQUARIUS: The Water Bearer

MAKEUP: 13 stars

BEST VIEWED: Sept./Oct.

LOCATION: Southwest

DEEP-SKY OBJECT:
M2, star cluster

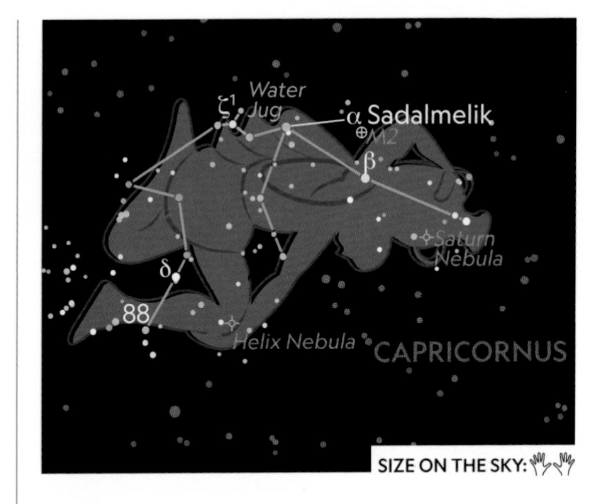

AMONG THE FAINTER CONSTELLATIONS of the zodiac, Aquarius lies between Pisces and Capricornus, south of the square of Pegasus. It is most visible in the fall months, when it reaches its highest point in the middle of the southern sky. The star 88 Aquari is an orange-red giant visible by the naked eye.

Two planetary nebulae are visible by telescope. The Helix Nebula is the closest to Earth, so despite its relatively low density it appears as large as half the diameter of the full moon. The Saturn Nebula is smaller and further away but when found it appears as a brilliant, green-tinted pinpoint.

Mythology

Many of this constellation's prominent stars have Arabic names that suggest the principles of good fortune. Sadalmelik, the alpha star, translates to "lucky one of the king"—an association consistent with Egyptian visions of this constellation bringing the yearly Nile River flood to their parched crops; the zodiacal symbol for the constellation is the hieroglyphic character for water. Aquarius is also seen as Ganymede, who was cupbearer to the god Zeus. Some see him as responsible for pouring water or wine from a jug, maintaining the mighty celestial river, the constellation Eridanus.

INTERESTING OBJECTS in Aquarius

Peaking on July 29, the annual Delta Aquarid meteor shower will whet sky-watchers' appetites for its more productive sister, the Perseids, which peaks a few weeks later. Best times to watch are in the predawn hours when its home constellation rises. Under clear, dark skies, expect 15 to 30 shooting stars per hour, with a trickle visible until mid-August.

CAPRICORNUS: The Sea Goat

MAKEUP: 12 stars

BEST VIEWED: Aug./Sept.

LOCATION: Southeast

DEEP-SKY OBJECT:
M30, globular star cluster

THE STARS THAT FORM this zodiacal constellation are not the brightest—magnitude 3 at most—but their broad, triangular shape is easily recognizable under a dark sky. The constellation lies southeast of bright Altair in Aquila and is bordered by Sagittarius, Piscis Austrinus, and Aquarius.

Lying about 27,000 light-years from Earth is pretty globular cluster M30. A challenge to resolve, its densely packed core will come into focus in small telescopes, but it can be spotted with binoculars. The cluster's retrograde motion through our galaxy suggests that it was cannibalized from a hapless small satellite galaxy that wandered too close and got sucked up by the Milky Way's enormous gravitational forces.

Mythology

Long recognized as a goat, Capricornus was subsequently given a fish tail. In one story, the god Pan leaped into the River Nile to escape a monster and the water transformed him. In an older tale, it represents one of Zeus's warriors in the battle with the Titans. He discovered conch shells, whose resounding call frightened the Titans into retreat. In appreciation, Zeus placed him in the sky with a fish tail and horns to represent his discovery.

M30 globular cluster

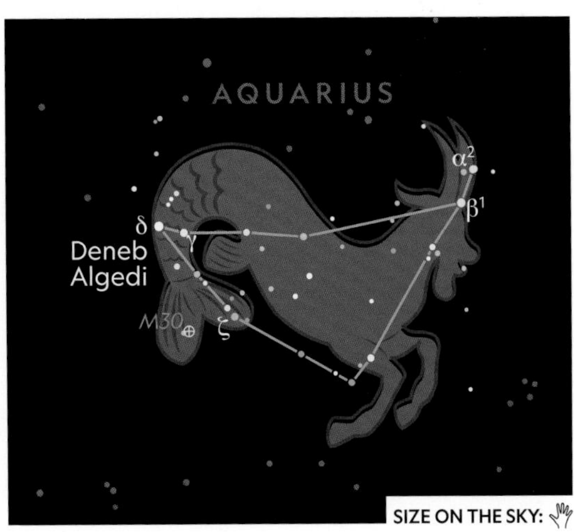

PISCES: The Fish

MAKEUP: 17 stars

BEST VIEWED: Oct./Nov.

LOCATION: Southeast

DEEP-SKY OBJECT:
M74, spiral galaxy

INTERESTING OBJECTS in Pisces

The westernmost star in the Circlet asterism, dubbed TX Piscium, is most famous for being a highly irregular variable. It is known as the reddest star visible, and it varies erratically in brightness from magnitude 4.5 to 6.2. This puts this 900-light-years-distant star within range of binoculars, and even the naked eye from a dark site when at its peak.

THIS ANCIENT CONSTELLATION of the zodiac is in the shape of two fish, their tails attached by a long cord. Pisces is located near the Square of Pegasus, overhead and to the southeast in October and November. The length of cord that holds the two fish together meets at alpha star Alrisha, just outside the constellation Cetus, making them form a large V in the sky.

Five faint stars in the south of the constellation form the asterism known as the Circlet, representing the head of the larger fish. Zeta Piscium (ζ) is a stunning double star shining at magnitudes 5 and 6. Van Maanen's star, at magnitude 12, is a rare white dwarf—the hot, naked core of a dead star—that is within reach of an 8-inch (200 mm) backyard telescope. A scope of that size will also resolve M74, a faint, face-on spiral galaxy whose arms are located just outside of the smaller fish. Three supernovae explosions have been spotted in this 30-million-light-years-distant galaxy, and that's just in the 21st century.

Mythology

The ancient Greeks believed the goddess Aphrodite and her son, Eros, changed themselves into fish to escape the sea monster Typhon (represented by Cetus), and the cord holds mother and son together as they swim. In Roman stories, the fish are Venus and Cupid.

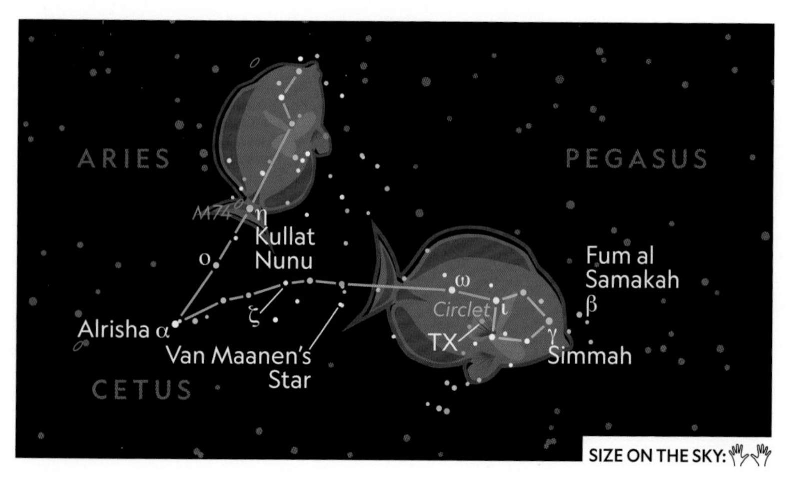

SIZE ON THE SKY:

DELPHINUS: The Dolphin

MAKEUP: 5 stars

BEST VIEWED: Aug./Sept.

LOCATION: Northeast

DEEP-SKY OBJECT:
Gamma Delphini, double star

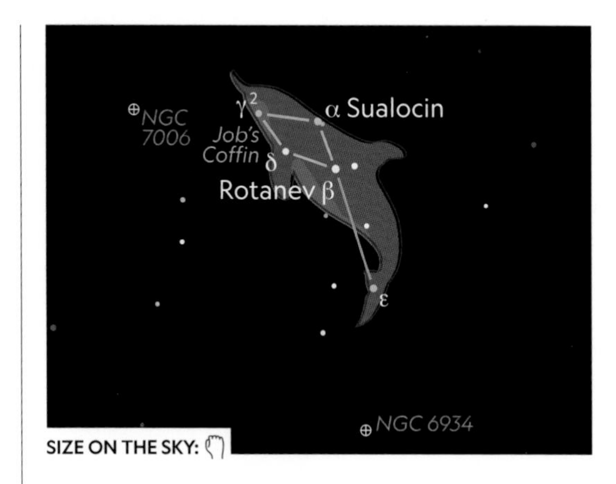

DELPHINUS MAKES UP for its humble size with a distinctive shape. Four stars form the asterism Job's Coffin—the body of the Dolphin—while one more represents its curved tail. This small constellation seems to swim toward Pegasus as it moves through the sky and can be found just west of a straight line traced between Altair in Aquila and Deneb in Cygnus.

In 1814, the alpha and beta stars in the group that make up the body of Delphinus were respectively named Sualocin and Rotanev in a star catalog published by the Palermo Observatory. When reversed, the letters in these star names spell Nicolaus Venator, the Latin form of Niccolo Cacciatore, who was the assistant director of the observatory at the time.

Delphinus's deep-sky object is one of its stars, Gamma Delphini (γ). About 101 light-years away, this optical double has a separation of 10 arc seconds. The pair is best seen through a telescope. The dimmer star has a slight green tinge.

Mythology

Delphinus was granted a place in the sky because he was a favorite of the Greek sea god Poseidon. The little dolphin is fabled to have convinced the Nereid Amphitrite to marry the god, who had been otherwise unsuccessful at attracting her attention.

INTERESTING OBJECTS in Delphinus

This tiny, kitelike constellation is home to two distant, beautiful globular clusters visible through small telescopes. Next to Epsilon Delphini, NGC 6934 is 50,000 light-years away. Near Gamma Delphini, NGC 7006 lies a whopping 137,000 light-years from Earth.

LACERTA: The Lizard

MAKEUP: 8 stars

BEST VIEWED: Oct.

LOCATION: Center of chart

DEEP-SKY OBJECT:
BL Lacertae

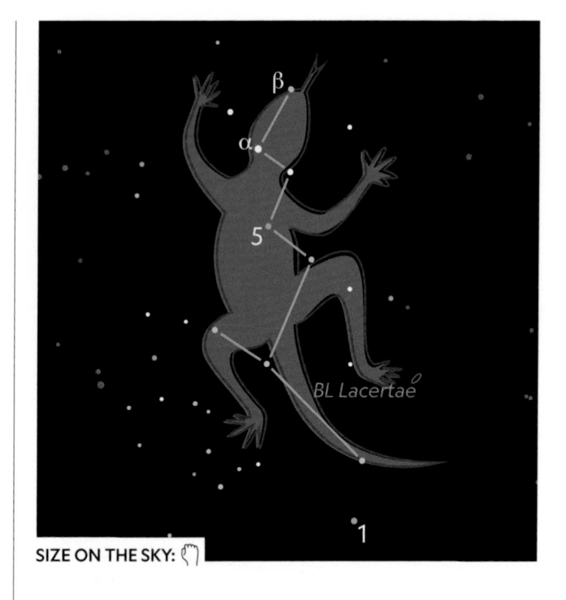

THIS NORTHERN CONSTELLATION was identified and named by Johannes Hevelius. Lacerta was first a small mammal, but soon it developed into a lizard. It is a group of fairly faint stars that occupies a small space in the sky, but its zigzag shape is easy to recognize on a dark night. It forms a zigzagging W, similar to the shape and size of neighboring Cassiopeia but is much dimmer. Lacerta can be found between Cygnus, Cassiopeia, and Andromeda. The brighter stars of its neighbors can overshadow Lacerta, making observation a challenge for the first-time viewer.

None of the stars in this constellation have merited proper names and the most distinct object, BL Lacertae, is only within reach of the most powerful amateur telescopes—but it is worth noting. The center of an elliptical galaxy, it varies in magnitude from 13 to 16.1, but it is not a classic variable star. Astronomers believe it is a type of quasar (see page 169) with a black hole at its core that jets a stream of light and energy in the direction of Earth. Lacerta contains no Messier objects, but it does have some open star clusters.

FURTHER

Because this constellation was so recently created to fill a gap in the sky left by the ancient charts, there is no relevant mythology or associated backstory. But there were competing efforts to name this piece of stellar turf, in honor of both Louis XIV and Frederick the Great.

TRIANGULUM: The Triangle

MAKEUP: 3 stars

BEST VIEWED: Nov./Dec.

LOCATION: Southeast

DEEP-SKY OBJECT:
M33, Pinwheel galaxy

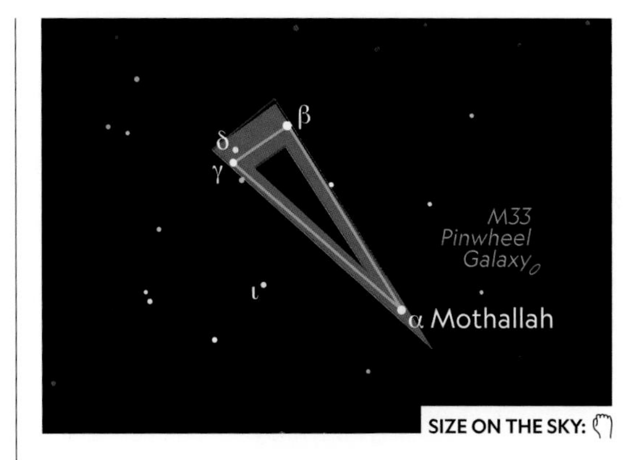

ONCE THESE RATHER FAINT STARS are identified just east of Andromeda, the Triangle is easily identified. It is one of Ptolemy's original 48 constellations and has many different associations across cultures. Before it took on its contemporary name, the ancient Greeks paid attention to this indistinct group of stars because it looked like their letter delta (Δ) and so, for some time, was known as Deltoton.

Stars & Objects

The most distinct object in Triangulum is the Pinwheel galaxy (M33), part of the local group of galaxies that includes the Milky Way. M33 is three million light-years away from Earth. Like our own galaxy, the Pinwheel is a spiral galaxy, its shape distinct when viewed from Earth's head-on perspective. The Pinwheel is bright enough to be spotted as a slight glow on dark nights with very good viewing conditions, but a telescope with a wide field of view is needed to get the full pinwheel effect.

In 2007 a black hole about 15.7 times the mass of the sun was discovered in the galaxy orbiting a companion star. Scientists predict that the black hole's partner star will eventually go supernova, and then there will be two black holes in the galaxy.

Pinwheel galaxy (M33)

CETUS: The Sea Monster

MAKEUP: 13 stars

BEST VIEWED: Nov.

LOCATION: Southeast

DEEP-SKY OBJECT: Mira, variable star

THIS VERY LARGE, FAINT CONSTELLATION has quite a distinctive shape. Its head forms a small group of stars connected to the body by Omicron Ceti, or Mira, the most famous long-period variable star. Watching 424-light-years-distant Mira can be a worthwhile project and a helpful way to develop your own sense of stellar magnitude. At its brightest, the star reaches around magnitude 2.4, but then over 11 months it fades to magnitude 9.3—invisible to the naked eye.

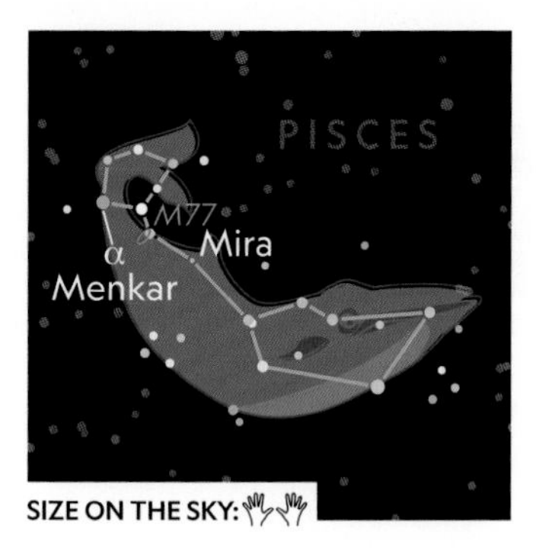

SIZE ON THE SKY:

Mythology

The sea god Poseidon sent Cetus, the giant whale or monster, to terrorize Ethiopia after Queen Cassiopeia boasted that her daughter was more beautiful than his sea nymphs, the Nereids. The constellation is also said to represent the whale that swallowed Jonah in the Old Testament.

EQUULEUS: The Little Horse

MAKEUP: 4 stars

BEST VIEWED: July/Aug.

LOCATION: Southeast

DEEP-SKY OBJECT: None

LOCATED IN the crowded southern portion of the sky, it's easy to overlook the Little Horse. It rests between much larger constellations: Pegasus, its equine counterpart to the northeast, and Aquila, the Eagle, directly southwest. Both make good signposts to help locate it.

Though one of the more inconsequential shapes in the sky, Equuleus's creation is attributed to one of astronomy's early greats—the Greek observer Hipparchus. Still, Ptolemy recognized the constellation as only a partial horse, while Arab astronomers thought the star Alpha Equulei—also known as Kitalpha—was the only significant feature of the constellation.

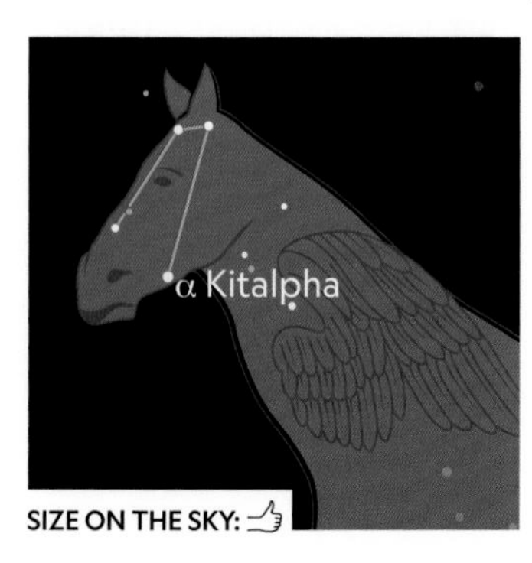

SIZE ON THE SKY:

Mythology

Equuleus is thought to represent Celeris, brother-horse of Pegasus in Greek mythology and said to have belonged to Castor, twin brother of Pollux.

PISCIS AUSTRINUS: The Southern Fish

MAKEUP: 11 stars

BEST VIEWED: Sept./Oct.

LOCATION: Southwest

DEEP-SKY OBJECT: None

THE ANCIENT EGYPTIANS SAW this group of stars as a fish. Piscis Austrinus is easily identified in the fall just south of Aquarius and east of Capricornus—its companions in this "watery" section of the sky. Fomalhaut is the most distinctive object in this constellation, the 18th brightest star to the naked eye and just 25 light-years from Earth. Fomalhaut is a relatively young star—between 100 and 300 million years old, with a life span estimated at a billion years. It is twice as large as the sun and burns between 14 and 17 times as bright. A breakthrough observation of the star was made in November 2008, when the Hubble telescope photographed what appeared to be a planet orbiting it. Scientists named the planet Fomalhaut b, but as they looked closer, doubts were raised about its makeup—it is shrouded in dust and unidentified light emissions.

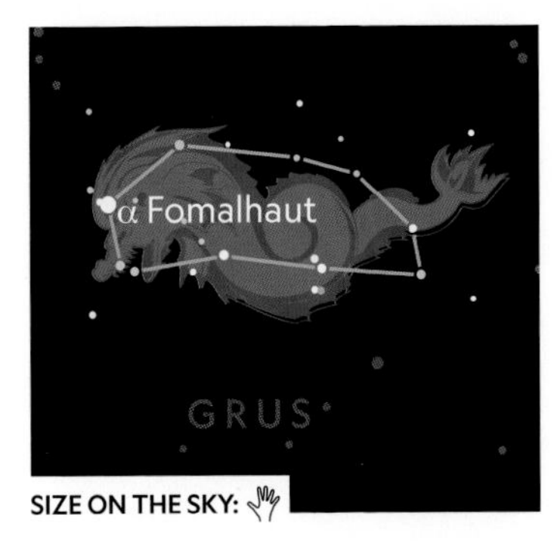

SIZE ON THE SKY:

GRUS: The Crane

MAKEUP: 11 stars

BEST VIEWED: Sept.

LOCATION: Southwest

DEEP-SKY OBJECT: Grus Quartet of galaxies

THE CRANE emerges toward the southern sky for a short time in the fall. It can be identified by its X shape with the red giant Beta Gruis (β) as its bright central star just south of Fomalhaut in Piscis Austrinus.

The three brightest stars in the constellation are unremarkable, but they show how apparent magnitude can be affected by distance. Alpha Gruis—also known as Alnair, meaning the "bright one of the tail" in Arabic—is classified as the brightest star partly because it is just 100 light-years away and is a large blue star. The second brightest star, Beta Gruis, is more than 10 times brighter in an absolute sense, but it ranks second because its appearance is dimmed by its distance from Earth: 170 light-years. Gamma Gruis is intrinsically brighter than both stars but is 200 light-years away. Within the Crane is the Grus Quartet, a group of four spiral galaxies: NGC 7552, 7582, 7590, and 7599.

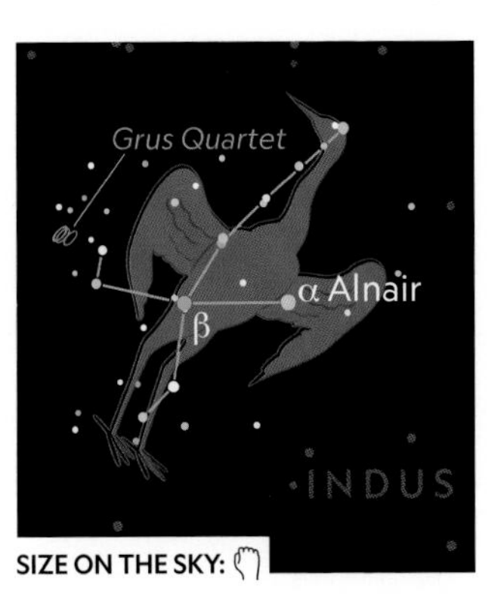

SIZE ON THE SKY:

WINTER

BRIGHT ORANGE BETELGEUSE sits on Orion's left shoulder, and blue-white Rigel forms the right foot of the mythical hunter. In the sword hanging from the hunter's belt of three stars, you'll find its middle star is in fact the Great Orion Nebula star factory. Just to the northwest of Betelgeuse is Taurus and the two famed open star clusters, the Pleiades and the Hyades.

DATE	TIME
12/21	11 p.m.
1/21	9 p.m.
2/1	8 p.m.

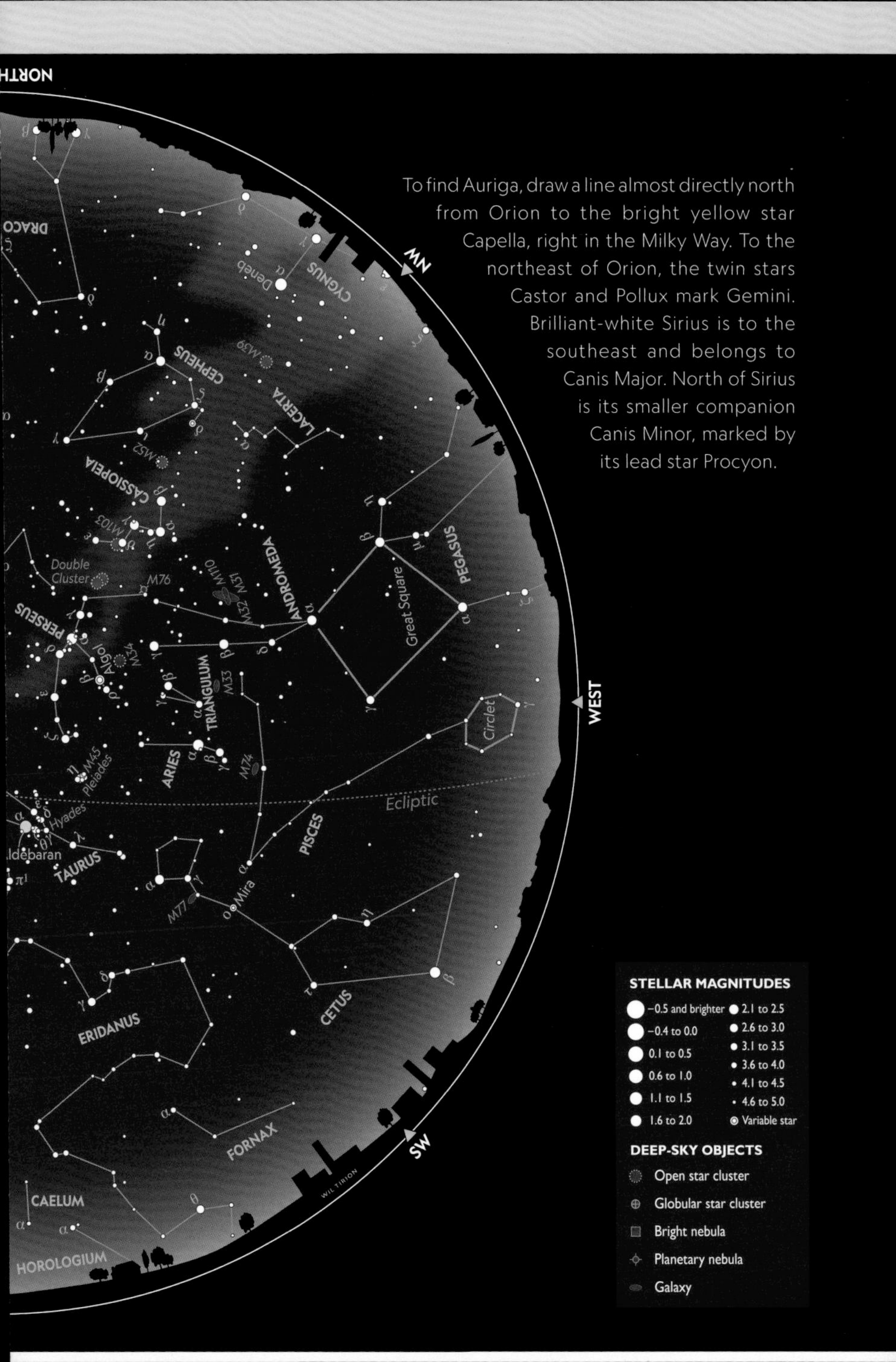

To find Auriga, draw a line almost directly north from Orion to the bright yellow star Capella, right in the Milky Way. To the northeast of Orion, the twin stars Castor and Pollux mark Gemini. Brilliant-white Sirius is to the southeast and belongs to Canis Major. North of Sirius is its smaller companion Canis Minor, marked by its lead star Procyon.

WINTER STAR HOPPING

LOW IN THE EAST, Leo—with its sickle-shaped pattern of stars—lies on its side. The Big Dipper of Ursa Major appears to be standing on its handle low in the northeastern sky. Extending imaginary lines from its "bowl" across the overhead sky leads to some of winter's brightest stars: brilliant-white Castor and Pollux in Gemini and yellow-hued Capella of Auriga. Dominating the southern sky is the biggest asterism of the season, the Winter Hexagon. We start with Capella, then hop clockwise to trace out its seven stars: find orange Aldebaran, then Rigel and Sirius, at which point moving up the eastern side takes us to Procyon, Pollux, and Castor, then back to Capella. Aldebaran is the brightest star in Taurus (the Bull) where Charles Messier sighted the first entry into his catalog: M1, the Crab Nebula.

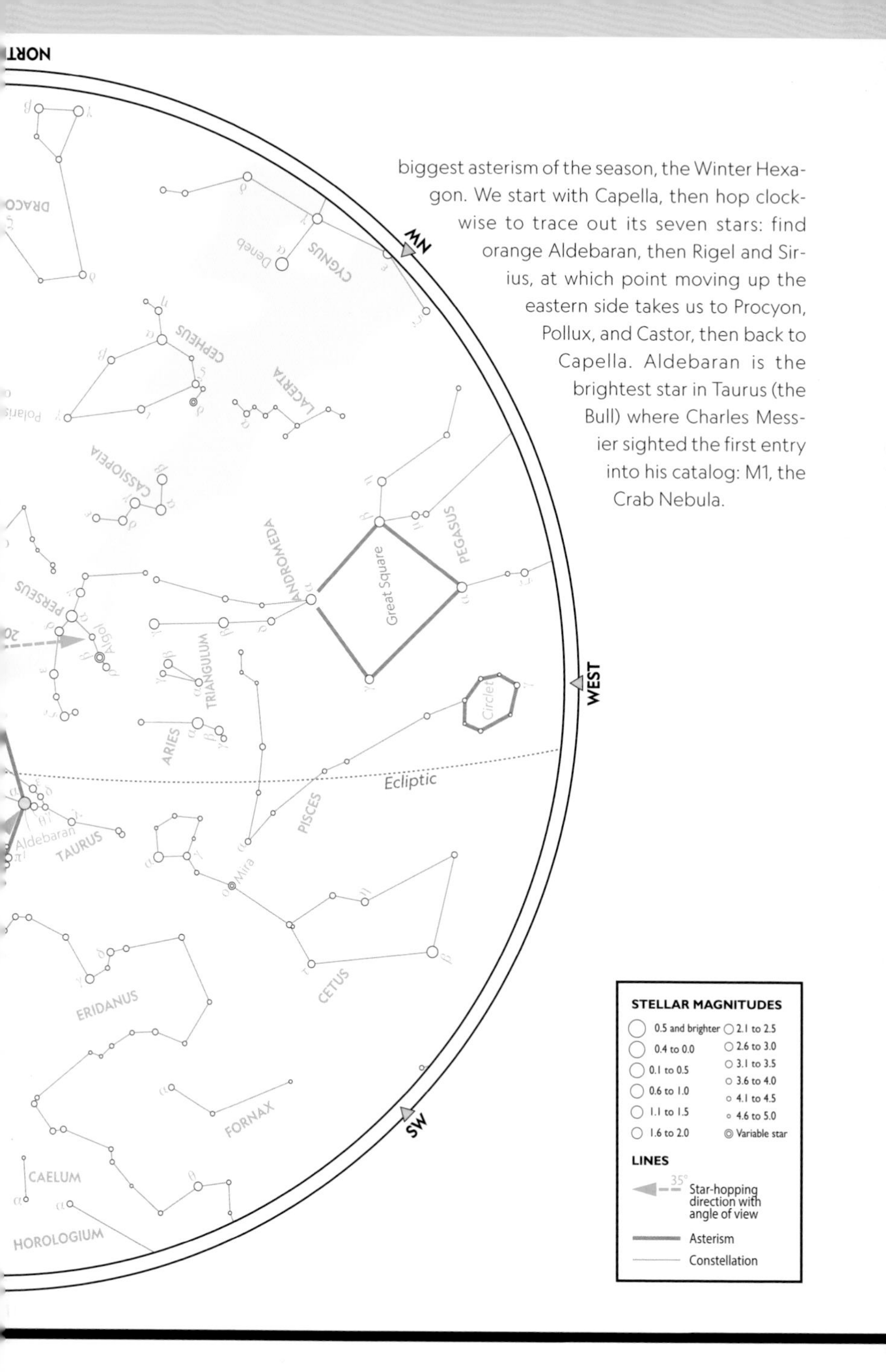

ORION: The Hunter

MAKEUP: 20 stars

BEST VIEWED: Jan./Feb.

LOCATION: Southeast

DEEP-SKY OBJECT: M42, Orion Nebula

THIS MAY BE THE MOST RECOGNIZABLE constellation in the sky, visible worldwide from its position straddling the celestial equator and identified by different cultures across thousands of years. Its prominence earned it three references in the Bible as well as mentions in *The Iliad, The Odyssey,* and other literary classics. To modern sky-watchers, the hunter's position is ideal for use as a reference point: Three of the sky's 25 brightest stars are found in this group, and their brightness makes the structure of a man quite easy to see. The hunter's bright belt is made of three stars, all around magnitude 2. His upraised, club-wielding arm sweeps into the Milky Way, while his other hand holds a shape interpreted as either a shield or a lion's pelt.

Bright Stars

Orion's main stars provide a way to orient yourself even

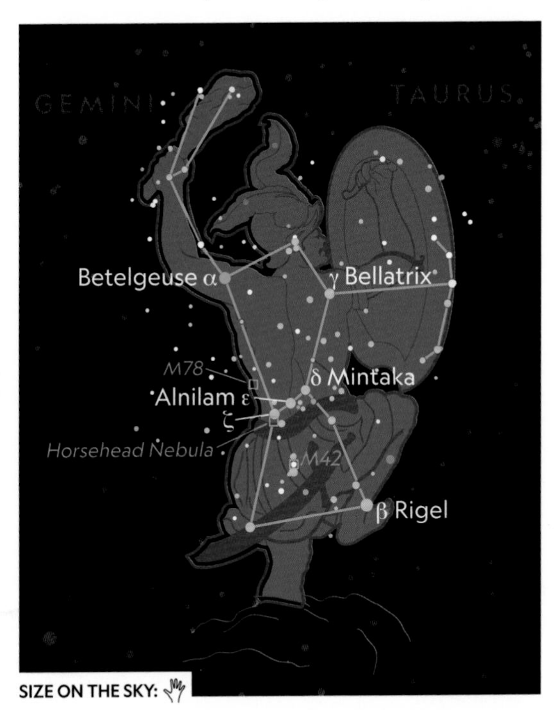

SIZE ON THE SKY:

FURTHER

If Betelgeuse replaced the sun at the center of our solar system it would extend to the orbit of Jupiter. This supergiant star is 10 million years old and near the end of its life, expected to explode as a supernova anytime—within the next million years.

Star formation in the Great Orion Nebula (M42)

amid the glare of city lights. Betelgeuse, the constellation's alpha star, represents one shoulder. This huge variable star pulsates to a diameter between 300 and 400 times the size of our sun. Blue-white Rigel, the hunter's foot, is a supergiant, 57,000 times as bright as the sun.

Mythology

The image is so intuitively that of a large person with a raised club that many different cultures around the would have given the group of stars similar interpretation. It was called Al Jabbar (the Giant) by Syrian astronomers and Sahu (the soul of the god Osiris) by the Egyptians. An ocean away, Native Americans in New Mexico referred to it as Long Sash, a heroic figure from their own mythology.

There are a few stories In Greco-Roman mythology associated with Orion. In one, Orion was a powerful hunter who was dealt a lethal sting by Scorpius—the reason his glowing foot is held high and why the two are placed, as they are, on opposite sides of the sky. In another, he is the love interest of Apollo, whose sister Artemis—the goddess of the hunt—chose Orion as her hunting companion and placed him in the sky after his premature passing.

INTERESTING OBJECTS in Orion

Orion includes an area of the Milky Way that features intense star production. Beneath the hunter's belt, in the middle of three stars that form Orion's sword, is the Great Orion Nebula (M42). Visible to the naked eye as a cloud patch, larger telescopes will show 1,300-light-years-distant gas clouds where star formation is taking place at a furious pace. At its center are four newborn stars known as the Trapezium.

MAIN STARS

β | RIGEL
COLOR: Blue
MAGNITUDE: 0.2
DISTANCE (LY): 864

α | BETELGEUSE
COLOR: Red
MAGNITUDE: 0.5
DISTANCE (LY): 642

γ | BELLATRIX
COLOR: Blue-white
MAGNITUDE: 1.6
DISTANCE (LY): 240

ε | ALNILAM
COLOR: Blue
MAGNITUDE: 1.7
DISTANCE (LY): 915

δ | MINTAKA
COLOR: Blue
MAGNITUDE: 2.2
DISTANCE (LY): 915

TAURUS: The Bull

MAKEUP: 13 stars

BEST VIEWED: Jan./Feb.

LOCATION: Southwest

DEEP-SKY OBJECT:
M45, Pleiades;
M1, Crab Nebula

TAURUS IS A CONSTELLATION of the zodiac at its northern peak in winter. It is easy to locate because Orion's belt points right toward Aldebaran, the red eye of the bull. Farther to the northeast, the 444-light-years-distant Pleiades star cluster marks the animal's shoulder. Also known as the Seven Sisters, it is one of the more easily identified deep-sky objects. A line traced from Betelgeuse (in Orion) through Aldebaran will bring the dazzling group into view. Myths of many cultures have been spun around the seven naked-eye stars in this cluster. A Native American story tells of seven young sisters who took a walk and lost their way. Now they stay in the sky as a reminder to children not

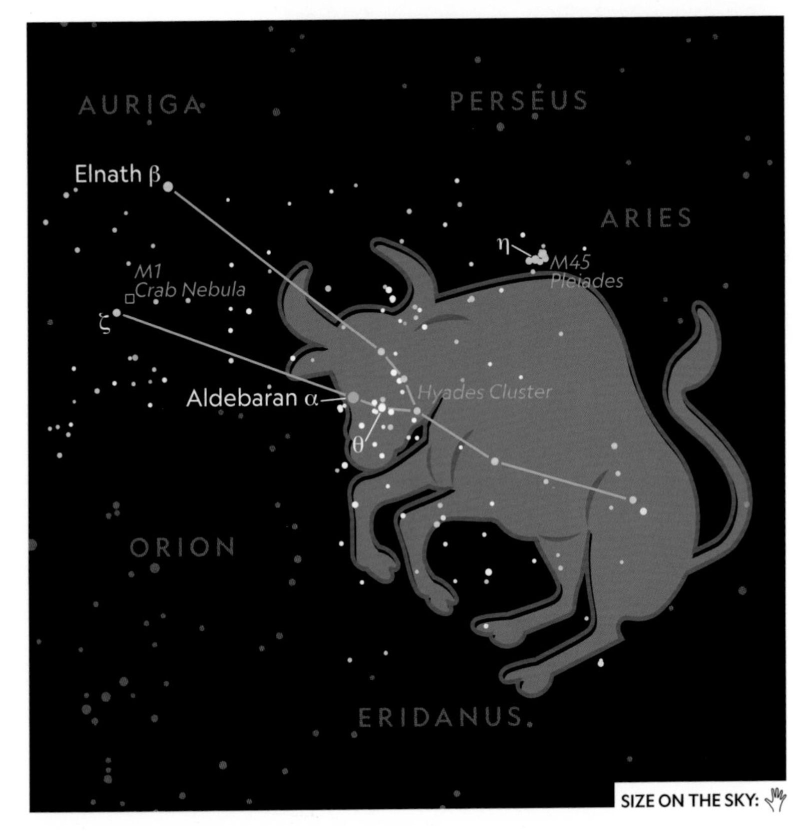

INTERESTING OBJECTS in Taurus

In 1758, the Crab Nebula prompted Charles Messier to begin his extensive catalog of deep-sky objects, earning it the title M1. The supernova that produced the Crab Nebula appears in records worldwide in A.D. 1054, from ancient Chinese writings to cave art in the American Southwest. The 6,500-light-years-distant supernova was said to have been visible to the naked eye for almost two years in the night sky, and visible during the day for almost a month. Now it is visible through small telescopes as a tiny gray cloud.

Crab Nebula is a star that exploded long ago.

to stray too far from home. Binoculars will help reveal the cluster's 500-plus stars.

Another interesting object to look for is the Hyades star cluster. This tight, V-shaped gathering of stars lies about 150 light-years from Earth, making it one of the closest clusters. While only about a dozen stars within the Hyades cluster can be spotted with the naked eye on a dark, moonless night, hundreds of members of this grouping are visible through binoculars.

Mythology

Numerous myths accompany this constellation. The ancient Egyptians associate the Bull with Osiris, god of life and fertility, while in Greek mythology the Bull is one of Zeus's disguises, used to capture Europa and bring her across the sea to the continent that now carries her name. To some, Taurus represents the golden calf formed by the idling followers of Moses while he received the Ten Commandments.

MAIN STARS

α | ALDEBARAN
COLOR: Orange
MAGNITUDE: 1.0
DISTANCE (LY): 67

β | ELNATH
COLOR: Blue-white
MAGNITUDE: 1.7
DISTANCE (LY): 134

η | ALCYONE
COLOR: Blue-white
MAGNITUDE: 2.9
DISTANCE (LY): 400

ζ | ZETA TAURI
COLOR: Blue-white
MAGNITUDE: 3.0
DISTANCE (LY): 440

θ | THETA TAURI
COLOR: Yellow-orange
MAGNITUDE: 3.8
DISTANCE (LY): 154

GEMINI: The Twins

MAKEUP: 13 stars

BEST VIEWED: Feb./Mar.

LOCATION: Southeast

DEEP-SKY OBJECT:
M35, star cluster

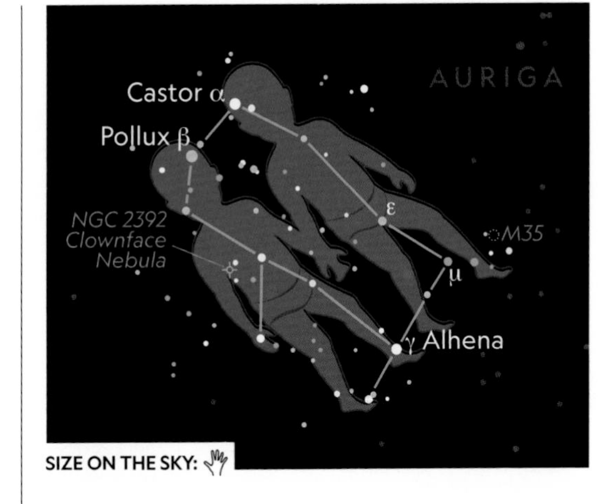

SIZE ON THE SKY:

OVERHEAD IN LATE WINTER, sky-watchers can see a group of stars long identified as twins—the famous brothers Castor and Pollux. The feet of the Gemini twins are just northeast of Betelgeuse, the bright star at the shoulder of Orion's upraised arm. At their heads are two giant stars, 50-light-years-distant Castor and 34-light-years-distant Pollux. Castor, the fainter of the twins, is a pretty double star that is actually a six-star system, three of which are visible with a small telescope. Pollux is the 17th brightest star in the night sky and nine times larger than our sun.

The annual Geminids is one of the more impressive meteor showers, radiating out around December 13th or 14th when the constellation rises in the east after midnight. Shooting stars streak by at a rate of 60 to 120 stars an hour, depending on sky conditions.

Mythology

The constellation's alpha and beta stars are named after the twins born of Leda, queen of Sparta, and Zeus in Greek mythology, but some legends have them as half-brothers. They served as shipmates with Jason on the ship Argo in his quest for the Golden Fleece. Blue-colored Castor and yellow-hued Pollux also had a sister named Helen, whose beauty instigated the Trojan War.

INTERESTING OBJECTS in Gemini

Messier 35 resides near the three "foot stars" of Gemini. This open star cluster consists of hundreds of relatively young stars about 100 million years old. The cluster sits some 2,800 light-years away and stretches 20 light-years across; in our Earthly skies, it takes up about the same amount of space as the full moon. While visible with the naked eye, the best views of this cluster are through binoculars and telescopes.

CANIS MAJOR: The Larger Dog

MAKEUP: 8 stars

BEST VIEWED: Jan./Feb.

LOCATION: Southeast

DEEP-SKY OBJECT: M41, Little Beehive star cluster

CANIS MAJOR IS LOCATED near the Milky Way, just to the east of Orion. The constellation is most easily spotted by locating its alpha star, Sirius—the brightest star in the night sky. Only 8.6 light-years away, it shines a brilliant magnitude of −1.5 and has a close companion white dwarf, though it's difficult to spot even with a 10-inch (250 mm) telescope. Sky-watchers in the south can identify Sirius by tracing a line through Orion's belt and continuing southeast, where the bright star marks the northern edge of Canis Major.

This constellation is home to several clusters and nebulae listed in the Messier and NGC lists. The brightest is open cluster M41, also known as the Little Beehive, which is about five degrees south of Sirius and sits 1,300 light-years distant. Easy to spot with binoculars as a hazy patch, M41 is particularly impressive when seen through a telescope, with orange stars nestled within.

Mythology

Canis Major is considered to be the larger of Orion's two hunting dogs. The phrase "dog days of summer" takes its origin from Sirius, known as the Dog Star. In late summer in the Northern Hemisphere, the star rises around the same time as the sun, leading to the belief that its heat and brightness help to bring extra warmth to the north.

Little Beehive star cluster (M41)

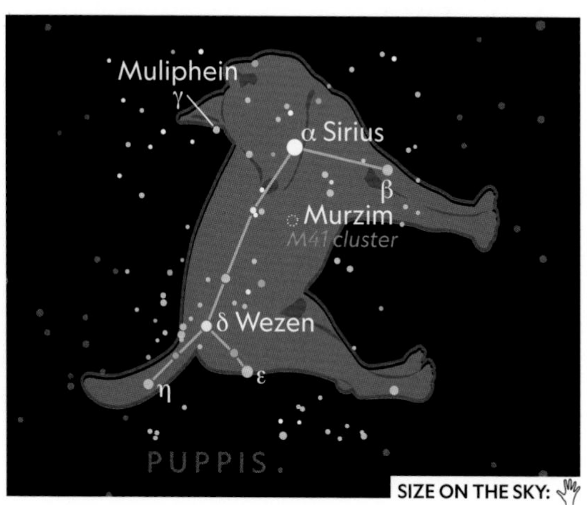

SIZE ON THE SKY:

CANIS MINOR: The Smaller Dog

MAKEUP: 2 stars

BEST VIEWED: Jan./Feb.

LOCATION: Southeast

DEEP-SKY OBJECT: None

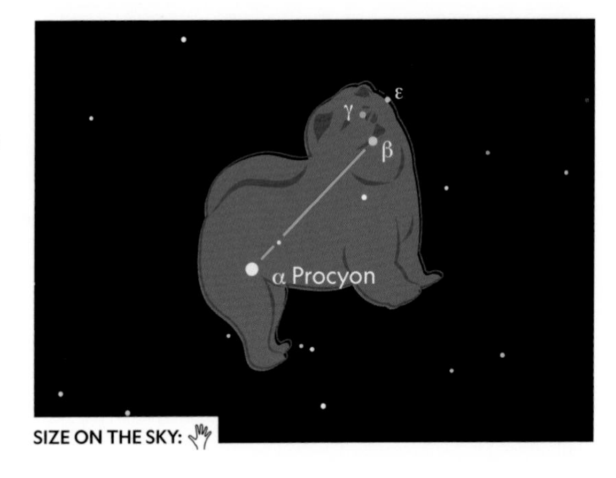

SIZE ON THE SKY:

ORION'S OTHER HUNTING DOG lies directly northeast of its larger companion, Canis Major. This smaller constellation (as the name implies) is not as bright or vibrant as its counterpart. Visualizing a young dog in this small set of stars takes some creativity, but the angle at which Canis Minor lies is evocative of the animal's cocked head as he looks toward his northern neighbor, Gemini. It appears sandwiched between the constellations of Cancer and Monoceros.

A challenge to find for the beginning backyard astronomer, this constellation is composed of only two stars brighter than magnitude 5 and contains no deep-sky objects brighter than magnitude 15. The alpha star, Procyon, is the brightest in Canis Minor and the eighth brightest star in the night sky. This deep-yellow star is only 11.4 light-years away from Earth and, like Sirius, also hosts its own white dwarf companion. It's 15,000 times fainter than Procyon, making it too faint to spot.

Mythology

Besides being considered one of Orion's hunting dogs, Canis Minor has other associations as well. Some legends have it resting under a table, expecting scraps from the Gemini twins Castor and Pollux. Another myth is that Canis Minor represents Helen of Troy's favorite dog, which allowed her to elope with Prince Paris.

FURTHER

Another seasonal signpost: The Winter Triangle asterism is formed by Canis Minor's Procyon, Orion's Betelgeuse, and Canis Major's Sirius. This asterism is the bright, seasonal counterpart to the Summer Triangle asterism formed by Altair, Deneb, and Vega.

ERIDANUS: The River

MAKEUP: 33 stars

BEST VIEWED: Dec./Jan.

LOCATION: Southwest

DEEP-SKY OBJECT: NGC 1300, spiral galaxy

STARGAZERS IN THE NORTHERN HEMISPHERE can follow the course of this winding constellation from its source just west of the foot of Orion until it cascades right over the horizon. Its bright alpha star, Achernar, lies at its southern tip but is not visible above roughly 20 degrees north latitude. Located 140 light-years away, Eridanus is the sky's sixth largest constellation and Achernar is the ninth brightest star. The stunningly blue star burns at very high temperatures, making it over 1,300 times more luminous than our sun. The star spins so fast on its axis that it's actually egg-shaped and is considered the galaxy's least round star.

Another star in Eridanus in the northern bend of the river is Epsilon Eridani, also called Ran, which is 10.5 light-years away and can be seen unaided. It is very similar to our sun, in having possible planets, and thus is being investigated for extraterrestrial life.

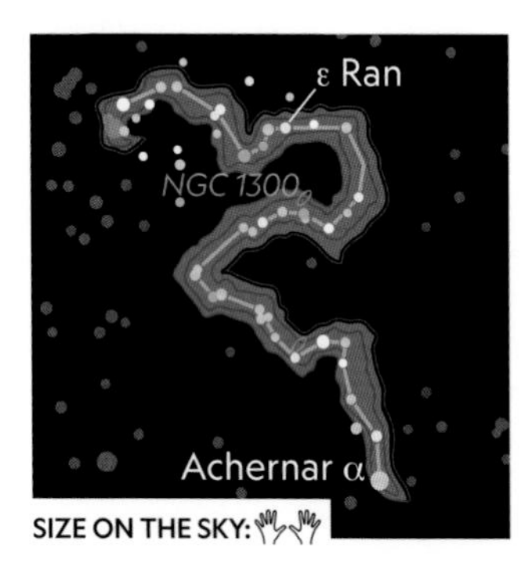

SIZE ON THE SKY:

MONOCEROS: The Unicorn

MAKEUP: 8 stars

BEST VIEWED: Jan./Feb.

LOCATION: Southeast

DEEP-SKY OBJECT: NGC 2244, Rosette Nebula

THIS CONSTELLATION WAS CREATED to fill gaps in the sky, suggested in 1624 by German astronomer Jakob Bartsch. Though dim, Monoceros fills a prominent gap in the sky bounded by very bright stars: Procyon in Canis Minor, Betelgeuse in Orion, and Sirius in Canis Major.

Stars & Objects

Though the Unicorn is faint, with no stars above magnitude 3.9, the Milky Way and its brighter neighbors offer many good reference points to locate it. The two stars closest to Betelgeuse represent the long head of the Unicorn. The open star cluster M50, nicknamed the Open-Heart cluster, is located 2,300 light-years away and can be glimpsed with binoculars, but it is best resolved with a small telescope. The shape of the large star-forming nebula NGC 2244 has been compared to both a rosette and a skull.

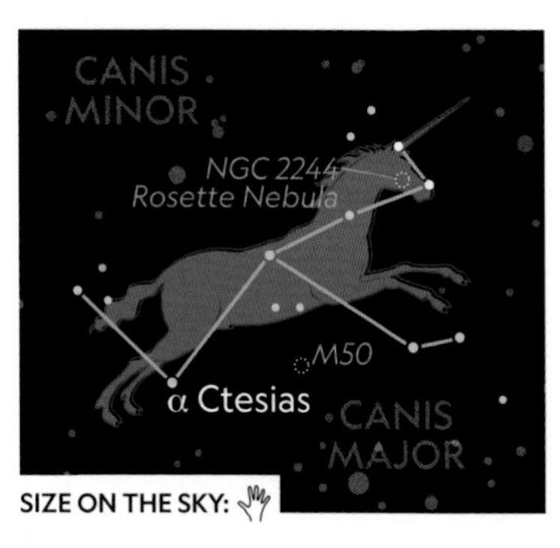

SIZE ON THE SKY:

PUPPIS: The Stern

MAKEUP: 12 stars

BEST VIEWED: Feb./Mar.

LOCATION: Southeast

DEEP-SKY OBJECT: M47, open cluster

THIS SOUTHERN CONSTELLATION FLOATS in the Heavenly Waters, an area that contains Eridanus, Pisces, Aquarius, and the three parts of the disassembled ship Argo that carried Jason and his Argonauts on their journey to steal the Golden Fleece. Its sail can be seen in Vela, its keel in Carina, and its anachronistic compass in Pyxis.

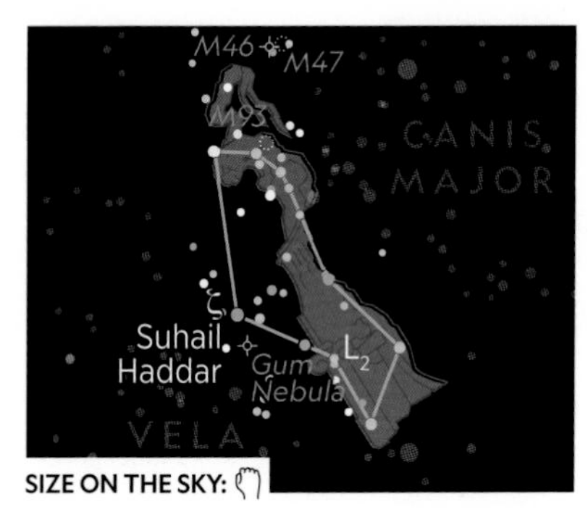

SIZE ON THE SKY:

Stars & Objects

Puppis contains no bright stars, but it is home to two faint open star clusters. M47 is easy to find with the naked eye under dark skies, looking like a hazy patch, and is home to only about 50 stars. Look closely through binoculars and you'll get a cosmic two-for-one-deal with a second open cluster, M46. A tad dimmer than its neighbor, it sits at a whopping 5,000 light-years away.

LEPUS: The Hare

MAKEUP: 11 stars

BEST VIEWED: Jan./Feb.

LOCATION: Southeast

DEEP-SKY OBJECT: Hind's Crimson Star, variable star

THIS SMALL CONSTELLATION is recognizable less for its shape than its position in the sky—the Hare's head is just below Rigel, Orion's blue-tinted foot. Its brightest star, Arneb, is magnitude 2.6. Hind's Crimson Star is a deep-red variable that ranges from a magnitude of 5.5 to a magnitude of 11.7 over roughly 14 months. It's an easy binocular target when at peak brightness, and its coloring is most striking. Visible with small telescopes, globular star cluster M79 lies 42,000 light-years away and is thought to have been gravitationally ripped from a nearby dwarf galaxy millions of years ago.

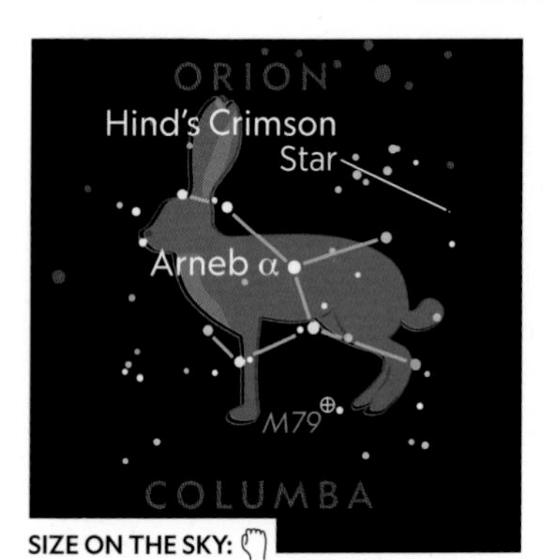

SIZE ON THE SKY:

Mythology

Ancient Egyptians referred to it as the boat of their god of the dead, Osiris, but it is commonly agreed that Orion and his hounds, Canis Major and Canis Minor, are hunting the hare through the sky.

COLUMBA: The Dove

MAKEUP: 8 stars

BEST VIEWED: Jan./Feb.

LOCATION: Southeast

DEEP-SKY OBJECT: NGC 1851, star cluster

THE DOVE IS ONE of the newer constellations, added by Dutch theologian and mapmaker Petrus Plancius in the 17th century to help fill out the sky.

The constellation honors the dove that acted as a scout during Noah's voyage on the ark. Its alpha star, Phact, has a magnitude of 2.6 and is about 270 light-years distant. Besides lying just south of Sirius, the brilliant star in Canis Major, Columba's distinguishing feature is the clear shape of three legs spiraling from a central star of medium brightness. It is located close to other, much brighter night sky objects that may outshine the stars of the Dove. A fairly large magnitude-7 globular star cluster, NGC 1851 appears as a hazy patch in binoculars at 9,200 light-years from Earth.

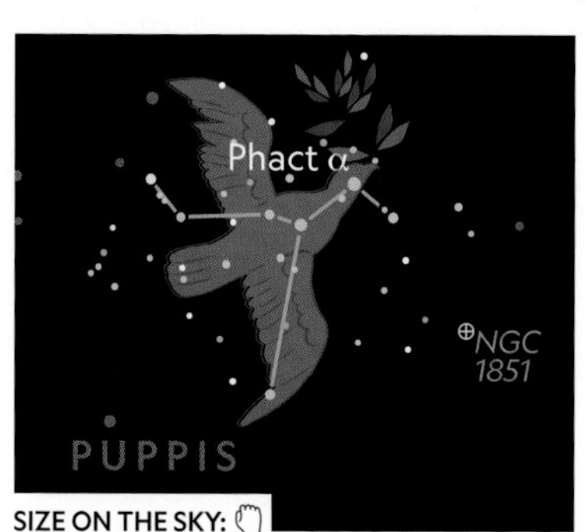

SIZE ON THE SKY:

LYNX: The Lynx

MAKEUP: 7 stars

BEST VIEWED: Feb./Mar.

LOCATION: Northeast

DEEP-SKY OBJECT: NGC 2419, Intergalactic Wanderer

NOT MANY CONSTELLATIONS are as inconspicuous as the Lynx. It is said that it earned its name because it requires the keen eyesight of a mountain cat to locate it.

Stars & Objects

Lynx's lone deep-sky object is the intriguingly named Intergalactic Wanderer (NGC 2419). Sitting some 274,000 light-years away, this aptly named globular cluster is more distant than satellite galaxies of the Milky Way and may one day even escape its gravitational pull and wander off into deep space. Because of its enormous distance—you'll need a telescope with a lens diameter of at least 10 inches (250 mm) to spot it under good viewing conditions—NGC 2419 appears as a tiny, fuzzy patch of light. It is technically within the Lynx's boundary but is actually closest to the star Castor.

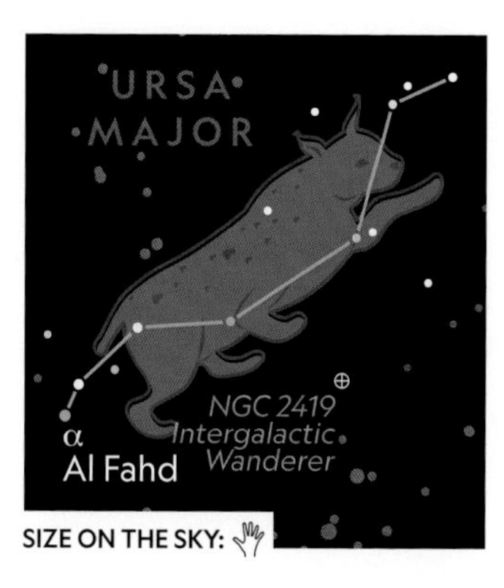

SIZE ON THE SKY:

| SOUTHERN STARS

SOME OF THE BEST celestial sights are visible only in the south. From 50 degrees south down to the South Pole, you'll be treated to glorious clusters, galaxies, nebulae, and stars. Australia makes a fine sky-watching site, with its clear, dry, dark skies. In April or May, Australia's autumn, the center of the Milky Way passes overhead, dense with stars. Below are just a few of the southern sky's finest sights.

Crux

A small, but distinct constellation, the Southern Cross (or, more formally, Crux) is an iconic sight in the southern sky and appears on the Australian flag. It is best seen from latitudes south of 20 degrees north. Wrapped on three sides by the constellation Centaurus, it includes several notable stars in its bright "T": Alpha Crucis, or Acrux, a double star at the foot of the cross; Gamma Crucis, or Gacrux, a double star at the northern end; and Beta Crucis, or Mimosa, which caps the cross's left arm. A gorgeous open cluster, the magnitude-4.2 Jewel Box gathers around the bright golden star Kappa Crucis, just below Beta Crucis. Through binoculars you can see this 6,400-light-years-distant cluster sparkle in different colors. Within Crux is a conspicuous dark patch just east of Alpha Crucis, called Coalsack Nebula. This dense cloud of gas and dust obscures the stars behind it.

Centaurus

This huge constellation is said to represent the centaur Chiron, who tutored Hercules. Located south of the constellation

The Very Large Telescopes at the Paranal Observatory have lenses that measure 5.9 feet (1.8 m) and can work in coordination or individually.

The bright stars in Crux are visible from an urban yard.

Hydra, it contains a wealth of notable objects. Among them are its brightest star, Alpha Centauri. This binary star (with Beta Centauri) is one of the sun's nearest neighbors, only 4.3 light-years away. A nearby companion visible only in telescopes, Proxima Centauri, is even closer, some 4.2 light-years distant. Centaurus also hosts the fifth brightest galaxy in the sky, Centaurus A, as well as the brightest, largest globular cluster in the sky, Omega Centauri (NGC 5139). Containing more than 10 million stars, this cluster is easily visible as a magnitude-3.9 large, fuzzy ball nearly as big as the full moon. It's best seen between April and September.

Large & Small Magellanic Clouds

These wispy, glowing patches are irregular dwarf galaxies that orbit the Milky Way. The Large Magellanic Cloud (LMC) is found in the constellation Dorado, not far from the south celestial pole; the Small Magellanic Cloud (SMC) floats within the constellation Tucana. The LMC is 168,000 light-years distant, while the SMC is about 200,000 light-years away, both much closer to Earth than the Andromeda galaxy.

Centaurus A galaxy

VIEWING THE NAKED-EYE PLANETS

The naked-eye planets can be seen without using binoculars or a telescope. These tables list the locations for the naked-eye planets through their annual journeys. With the exception of Mercury, the other planets will appear in constellations for months at a time during the year. At opposition, the planets will be the closest to Earth and will appear brighter than at any other time that year, offering the best chance to study details or take photographs. The approximate dates for opposition were predicted by *seasky.org*.

Mercury

Mercury can be found near the western horizon at dusk and the eastern horizon at dawn for several two-week intervals throughout the year. The peak dates were predicted by *seasky.org*.

YEAR	AT DUSK IN THE WEST	AT DAWN IN THE EAST
2019	February 27, June 23, October 20	April 11, August 9, November 28
2020	February 10, June 4, October 1	March 24, July 22, November 10
2021	January 24, May 17, September 14	March 6, July 4, October 25
2022	January 7, April 29, August 27, December 21	February 16, June 16, October 8
2023	April 11, August 10, December 4	January 30, May 29, September 22
2024	March 24, July 22, November 16	January 12, May 9, September 5, December 25
2025	March 8, July 4, October 29	April 21, August 19, December 7
2026	February 19, June 15, October 12	April 3, August 2, November 20
2027	February 3, May 28, September 24	March 17, July 15, November 4
2028	January 17, May 9, September 6, December 31	February 27, June 26, October 17
2029	April 21, August 19, December 14	February 9, June 8, October 1
2030	April 4, August 2, November 26	January 22, May 21, September 15

Transits of Mercury Across the Sun

DATE	UNIVERSAL TIME	SEPARATION*
November 11, 2019	15:20	76"
November 13, 2032	8:54	572"
November 7, 2039	8:46	822"
May 7, 2049	14:24	512"

Transits of Venus Across the Sun

DATE	UNIVERSAL TIME	SEPARATION*
December 11, 2117	2:48	724"

* Separation: distance (arc-seconds) between the centers of the sun and Mercury. Transit predictions by Fred Espenak, NASA's Goddard Space Flight Center

Venus

Boldface type indicates that Venus will be at its highest point above the horizon, predicted by *seasky.org*. *Italic type* indicates that the planet is not visible because it is too close to the sun.

	JAN.	FEB.	MAR.	APR.	MAY	JUNE	JULY	AUG.	SEPT.	OCT.	NOV.	DEC.
2019	**Lib**	Sgr	Cap	Aqr	Psc	Ari	Tau	*Cnc*	*Leo*	Vir	Sco	Sgr
2020	Cap	Aqr	**Psc**	Tau	Tau	*Tau*	Tau	**Tau**	Gem	Leo	Vir	Lib
2021	Oph	Cap	***Aqr***	*Psc*	Ari	Tau	Cnc	Leo	Vir	Lib	Sgr	Sgr
2022	Sgr	Sgr	Sgr	Cap	Psc	Ari	Tau	Gem	Leo	*Vir*	*Lib*	Oph
2023	Sgr	Aqr	Psc	Ari	Tau	**Gem**	Leo	Leo	Cnc	**Leo**	Leo	Vir
2024	Sco	Sgr	Cap	Psc	Ari	*Tau*	Gem	Leo	Vir	Lib	Oph	Sgr
2025	**Aqr**	Psc	Psc	Psc	**Psc**	Psc	Tau	Gem	Cnc	Leo	Vir	Lib
2026	*Sgr*	*Cap*	Aqr	Ari	Tau	Gem	Leo	**Vir**	Vir	Vir	Vir	Vir
2027	**Lib**	Sgr	Cap	Aqr	Psc	Ari	Tau	*Cnc*	*Leo*	Vir	Sco	Sgr
2028	Cap	Aqr	**Psc**	Tau	Tau	*Tau*	Tau	**Tau**	Gem	Leo	Vir	Lib
2029	Oph	Cap	*Aqr*	*Psc*	Ari	Tau	Cnc	Leo	Vir	**Lib**	Oph	Sgr
2030	*Sgr*	Sgr	**Sgr**	Cap	Psc	Ari	Tau	Gem	Leo	*Vir*	*Lib*	Oph

Mars

Boldface type indicates that the planet is at opposition.

	JAN.	FEB.	MAR.	APR.	MAY	JUNE	JULY	AUG.	SEPT.	OCT.	NOV.	DEC.
2019	Psc	Psc	Ari	Tau	Tau	Gem	Cnc	Leo	Leo	Vir	Vir	Lib
2020	Lib	Oph	Sgr	Cap	Cap	Aqr	Psc	Psc	Psc	**Psc**	Psc	Psc
2021	Psc	Ari	Tau	Tau	Gem	Gem	Cnc	Leo	Leo	Vir	Vir	Lib
2022	Oph	Sgr	Sgr	Cap	Aqr	Psc	Psc	Ari	Tau	Tau	Tau	**Tau**
2023	Tau	Tau	Tau	Gem	Gem	Cnc	Leo	Leo	Vir	Vir	Lib	Sco
2024	Sgr	Sgr	Cap	Aqr	Psc	Psc	Ari	Tau	Tau	Gem	Cnc	Cnc
2025	**Cnc**	Gem	Gem	Gem	Cnc	Leo	Leo	Vir	Vir	Vir	Lib	Oph
2026	Sgr	Cap	Aqr	Aqr	Psc	Ari	Tau	Tau	Gem	Cnc	Leo	Leo
2027	Leo	**Leo**	Leo	Leo	Leo	Leo	Leo	Vir	Vir	Lib	Oph	Sgr
2028	Sgr	Cap	Aqr	Psc	Ari	Tau	Tau	Gem	Cnc	Cnc	Leo	Leo
2029	Vir	Vir	**Vir**	Vir	Vir	Vir	Vir	Vir	Lib	Sco	Sgr	Sgr
2030	Cap	Aqr	Psc	Psc	Ari	Tau	Tau	Gem	Cnc	Leo	Leo	Vir

Jupiter

Boldface type indicates that the planet is at opposition.

	JAN.	FEB.	MAR.	APR.	MAY	JUNE	JULY	AUG.	SEPT.	OCT.	NOV.	DEC.
2019	Oph	Oph	Oph	Oph	Oph	**Oph**	Oph	Oph	Oph	Oph	Oph	Sgr
2020	Sgr	Sgr	Sgr	Sgr	Sgr	Sgr	**Sgr**	Sgr	Sgr	Sgr	Sgr	Sgr
2021	Cap	Cap	Cap	Cap	Aqr	Aqr	Aqr	**Aqr**	Cap	Cap	Cap	Cap
2022	Aqr	Aqr	Aqr	Aqr	Psc	Psc	Cet	Cet	**Psc**	Psc	Psc	Psc
2023	Psc	Psc	Psc	Psc	Psc	Ari	Ari	Ari	Ari	Ari	**Ari**	Ari
2024	Ari	Ari	Ari	Ari	Tau	Tau	Tau	Tau	Tau	Tau	Tau	**Tau**
2025	Tau	Tau	Tau	Tau	Tau	Tau	Gem	Gem	Gem	Gem	Gem	Gem
2026	**Gem**	Gem	Gem	Gem	Gem	Gem	Cnc	Cnc	Cnc	Leo	Leo	Leo
2027	Leo	**Leo**	Leo	Cnc	Cnc	Leo	Leo	Leo	Leo	Leo	Leo	Vir
2028	Vir	Vir	**Vir**	Leo	Leo	Leo	Leo	Vir	Vir	Vir	Vir	Vir
2029	Vir	Vir	Vir	**Vir**	Vir	Vir	Vir	Vir	Vir	Vir	Lib	Lib
2030	Lib	Lib	Lib	Lib	**Lib**	Lib	Lib	Lib	Lib	Lib	Lib	Oph

Saturn

Boldface type indicates that the planet is at opposition.

	JAN.	FEB.	MAR.	APR.	MAY	JUNE	JULY	AUG.	SEPT.	OCT.	NOV.	DEC.
2019	Sgr	Sgr	Sgr	Sgr	Sgr	Sgr	**Sgr**	Sgr	Sgr	Sgr	Sgr	Sgr
2020	Sgr	Sgr	Sgr	Cap	Cap	Cap	**Cap**	Sgr	Sgr	Sgr	Sgr	Sgr
2021	Cap	Cap	Cap	Cap	Cap	Cap	Cap	**Cap**	Cap	Cap	Cap	Cap
2022	Cap	Cap	Cap	Cap	Cap	Cap	Cap	**Cap**	Cap	Cap	Cap	Cap
2023	Cap	Cap	Aqr	Aqr	Aqr	Aqr	Aqr	**Aqr**	Aqr	Aqr	Aqr	Aqr
2024	Aqr	Aqr	Aqr	Aqr	Aqr	Aqr	Aqr	Aqr	**Aqr**	Aqr	Aqr	Aqr
2025	Aqr	Aqr	Aqr	Aqr	Psc	Psc	Psc	Psc	**Psc**	Aqr	Aqr	Aqr
2026	Aqr	Psc	Psc	Psc	Cet	Cet	Psc	Psc	Psc	**Cet**	Cet	Cet
2027	Cet	Cet	Psc	Psc	Psc	Psc	Psc	Psc	Psc	**Psc**	Psc	Psc
2028	Psc	Psc	Psc	Psc	Psc	Ari	Ari	Ari	Ari	**Ari**	Ari	Ari
2029	Ari	Ari	Ari	Ari	Ari	Ari	Ari	Tau	Tau	Tau	**Ari**	Ari
2030	Ari	Ari	Ari	Ari	Tau	Tau	Tau	Tau	Tau	Tau	**Tau**	Tau

FURTHER RESOURCES

Books

Aguilar, David A. *Space Encyclopedia: A Tour of Our Solar System and Beyond*. National Geographic Kids. Washington, D.C.: National Geographic, 2013.

Devorkin, David H., and Robert W. Smith. *The Hubble Cosmos: 25 Years of New Vistas in Space*. Washington, D.C.: National Geographic, 2015.

Devorkin, David H., and Robert W. Smith. *Hubble: Imaging Time and Space*. Washington, D.C.: National Geographic, 2008.

Dickinson, Terence. *Night Watch: A Practical Guide to Viewing the Universe*. Buffalo: Firefly Books, 2006.

Fazekas, Andrew. *Star Trek: The Official Guide to Our Universe*. Washington, D.C.: National Geographic, 2016.

Lang, Kenneth R. *The Cambridge Guide to the Solar System*, 2nd ed. Cambridge, UK: Cambridge University Press, 2011.

O'Meara, Stephen James. *Deep-Sky Companions: The Messier Objects*. Cambridge, UK: Cambridge University Press, 1998.

Trefil, James. *Space Atlas: Mapping the Universe and Beyond*, 2nd ed. Washington, D.C.: National Geographic, 2018.

Tyson, Neil deGrasse. *StarTalk: Everything You Ever Need to Know About Space Travel, Sci-Fi, the Human Race, the Universe, and Beyond*. Washington, D.C.: National Geographic, 2016.

The Royal Astronomical Society of Canada. *Observer's Handbook* (published annually).

Websites

GENERAL RESOURCES

Ask an Astronomer
Site run by Cornell University where users can post questions for a professional astronomer, *curious.astro.cornell.edu*

Astronomy Picture of the Day
A daily image or photograph explained by a professional astronomer, *apod.nasa.gov/apod/astropix.html*

Constellation Guide
Detail on constellations, their stars, and what to look for each month, *constellation-guide.com*

Kids Astronomy,
Free Astronomy website designed to teach children about the universe, *kidsastronomy.com*

Sky Maps
Free printable monthly all-sky maps with sky event calendars for both Northern and Southern Hemispheres, *skymaps.com*

Space
Multimedia website dedicated to space news, *space.com*

Space Weather
Information about the environment between the sun and Earth, *spaceweather.com*

NATIONAL & INTERNATIONAL ORGANIZATIONS

American Association of Variable Star Observers, *aavso.org*

American Meteor Society, *amsmeteors.org*

International Meteor Organization, *imo.net*

National Aeronautics and Space Administration, *nasa.gov*

Space Weather Prediction Center, *swpc.noaa.gov*

PERIODICALS

Astronomy magazine, *www.astronomy.com*

Sky & Telescope magazine, *skyandtelescope.com*

OBSERVATORIES & TELESCOPES

Chandra X-Ray Observatory, the first space-based x-ray observatory, *chandra.harvard.edu*

European Southern Observatory, *eso.org/public*

Hubble Site, website for NASA's Hubble Space Telescope, *hubblesite.org/newscenter*

National Optical Astronomy Observatory, *noao.edu*

Mauna Kea Observatories, *ifa.hawaii.edu/mko*

Palomar Observatory, *astro.caltech.edu/palomar/*

W. M. Keck Observatory, *keckobservatory.org*

Stargazing Apps and Software

StarChart

SkySafari, *skysafariastronomy.com*

StarryNight, *starrynighteducation.com*

Stellarium, *stellarium.org*

Planetariums & Museums

NORTHEAST AND SOUTHEAST
Albert Einstein Planetarium, Smithsonian National Air and Space Museum, Washington, D.C.

Fels Planetarium, the Franklin Institute, Philadelphia, Pennsylvania

Hayden Planetarium, American Museum of Natural History, New York, New York

Patricia and Phillip Frost Museum of Science, Miami, Florida

Saunders Planetarium, Museum of Science and Industry, Tampa, Florida

MIDWEST
Adler Planetarium, Chicago, Illinois

Nobel Planetarium, Fort Worth Museum of Science and History, Fort Worth, Texas

NORTHWEST AND SOUTHWEST
Aerospace Museum of California, McClellan Park, California

Dorrance Planetarium, Arizona Science Center, Phoenix, Arizona

Exploratorium, San Francisco, California

Oregon Museum of Science and Industry, Portland, Oregon

Samuel Oschin Planetarium-Griffith Observatory, Los Angeles, California

William Knox Holt Planetarium, University of California, Berkeley, California

ABOUT THE AUTHOR

ANDREW FAZEKAS, also known as the Night Sky Guy, is a science writer, speaker, and broadcaster who shares his passion for the wonders of the universe through all media. He writes the popular online column "StarStruck" for National Geographic and is the author of the National Geographic book *Star Trek: The Official Guide to Our Universe*. Fazekas is a syndicated correspondent for television and radio broadcast networks, the communications manager for Astronomers Without Borders, and an active sky-watching member of the Royal Astronomical Society of Canada since 1983. Co-creator of the world's first open-air augmented-reality planetarium experience in Canada, Fazekas and his team are now partnering with National Geographic to expand this groundbreaking science edutainment concept globally. He has never met a clear night sky he didn't like.

Howard Schneider, Patricia Daniels, John Scalzi, and Alan Dyer were the authors of the first edition of this book.

INDEX

N

O

P

Q

R

T

U

V

W

Y

Z

ILLUSTRATIONS CREDITS

T=Top, C=Center, B=Bottom, L=Left, R=Right, I=Inset

2MASS: 269 (L) Atlas Image obtained as part of the Two Micron All Sky Survey (2MASS), a joint project of UMass and the Infrared Processing and Analysis Center/CIT, funded by NASA and the NSF; **Alamy Stock Photo:** 27 Dominic Hatcher, 36 David Cheshire, 57 Carol Barrington; **AP Images:** 139 Alan Dyer/VW Pics, 180 Wolfgang Kumm/picture-alliance/dpa; **Patrick J. Bagley:** 20; **Tristan Brand:** Back cover (C); **Canon USA, Inc.:** 135 (B); **Celestron:** 134 (T), 135 (t); **Magnus Danbolt:** 21 (T); **Dreamstime:** 18 Aleksandra Alekseeva, 115 Dmitry Bomshtein; **Alan Dyer:** 118 (T), 118 (B), 138; **Ivan Eder, astroeder.com:** 197; **ESO:** 89, 120-121 S. Deiries, 165 Wolfram Freudling et al. (STECF), ESO, ESA, NASA, 169 M. Kornmesser, 233, 274 & Back cover (BR); **Fotosearch:** 176 (I) & 178 Visual Language; **Getty Images:** 8-9 Buena Vista Images, 15 QAI Publishing/UIG, 19 Allexxandar, 34 National Railway Museum/SSPL; 40-41 coolkengzz, 44 Paul Harris, 46-7 Picavet, 58 Justin Sullivan; 62 (T) NASA/SSPL, 64 Sjoerd van der Wal, 129 Universal History Archive, 172 Garth Eliassen; **Illustris Collaboration:** 171; **Henryk Kowalewski:** 243; **Tamas Ladanyi (TWAN):** 16; **Liverpool Telescope:** 245 Göran Nilsson & Wim van Berlo; **Meade Instruments:** 137; **Horst Meyerdierks:** 119 (B); **NASA:** 30-31, 45, 63 GSFC/SDO, 66-7 GSFC/ASU, 70 JPL, 74-5, 75 Goddard/ASU, 82 Bill Ingalls, 83 Terry Virts, 84-5 JPL-Caltech/SwRI/MSSS, 86, 88 GSFC/Reto Stöckli, 92 JHUAPL/Carnegie Institution of Washington, 94 JPL, 96, 99 JPL-Caltech/MSSS, 100 JPL-Caltech/UCLA/MPS/DLR/IDA, 101 JPL-Caltech/UCLA/MPS/DLR/IDA, 102 (L) STScI, 105 JPL-Caltech/SETI Institute, 106 Hubble Heritage Team (STScI/AURA), 109 (T) JPL-Caltech/ASI, 109 (B) JPL/SSI, 112 JHUAPL/SwRI, 113 JHUAPL/SwRI, 124 JPL-Caltech, 150 CXC/M.Weiss; **NASA/ESA:** 4 Digitized Sky Survey 2, 10 P. Kalas (UC Berkeley), 48 SOHO/EIT, 53 SOHO, 54 SOHO, 98 Z. Levay (STScI), 110 (L) L. Sromovsky and P. Fry (U of Wisconsin), H. Hammel (SSI), and K. Rages (SETI Institute), 111 (L) E. Karkoschka (U of Arizona), and H.B. Hammel (SSI), 140-41 Hubble Heritage Team, 143 Hubble Heritage Team (AURA/STScI), 149 Hubble Heritage Team (STScI/AURA); Ack: R. Gendler, 152 (L) J. Hester and A. Loll (ASU), 155 Hubble Heritage Team (STScI/AURA); Ack: C. Bailyn (Yale), W. Lewin (MIT), A. Sarajedini (U of Florida), and W. van Altena (Yale), 156 & Back cover (BL), STScI, 157 C. R. O'Dell (Vanderbilt), and M. Meixner, P. McCullough, and G. Bacon (STScI), 160-61 Hubble Heritage (STScI/AURA)/Hubble Collaboration, 162 CFHT, CXO, M.J. Jee (UC Davis), and A. Mahdavi (San Francisco State), 163, 168 B. Mobasher (STScI and ESA), 170 (TL) Hubble Heritage (STScI/AURA)/Hubble Collaboration. Ack: M. West (ESO, Chile), 170 (TR) Hubble Image: NASA, ESA, K. Kuntz (JHU), F. Bresolin (U. of Hawaii), J. Trauger (JPL), J. Mould (NOAO), Y.-H. Chu (U. of Illinois, Urbana), and STScI; CFHT Image: Canada-France-Hawaii Telescope/J.-C. Cuillandre/Coelum; NOAO Image: G. Jacoby, B. Bohannan, M. Hanna/NOAO/AURA/NSF, 170 (BL) Hubble Heritage Team (STScI/AURA), 198 (B) Hubble SM4 ERO Team, 201 STScI and C. Sarazin (U. of Virginia), 209 (B) S. Smartt (The Queen's University of Belfast), 218 © Robert O'Dell (Vanderbilt) & David Thompson (LBTO), 235 (B) Hubble, 247, 249 STScI and D. Geisler (U. de Concepción), 253 (L), 265 M. Robberto (STScI/ESA) and the Hubble Space Telescope Orion Treasury Project Team, 267 Ack: M. Weisskopf/MSFC; **NG Creative:** Cover & 35 (B) Babak Tafreshi, 32, Davis Meltzer, 145, Helmut K. Wimmer; **NG Maps:** 12-13 Manuel Canales & Matt Chwastyk, NGM Staff; Sean McNaughton. Art: Antoine Collignon. Sources: Guillermo Gonzalez, Ball State; Michael Gowanlock, Northern Arizona; Icarus; Astrobiology; NASA/JPL; International Journal of Astrobiology, 21 (B), 72-3, 76-7, 108, 132-3 Matt Chwastyk; **NG Maps/Debbie Gibbons:** 188, 194-5, 214-15, 240-41, 262-3; **NSF:** 166, N.R. Fuller; **Nikon, Inc.:** Melville, NY: 134 (B); **Orion:** 136; **Wally Pacholka:** 2-3; **Damian Peach:** 104; **Danilo Pivato:** 56; **Science Source:** 14 Larry Landolfi, 17 Doug Plummer, 22 Royal Astronomical Society, 24 Mark Garlick, 35 (T) Stephen & Donna O'Meara, 37 RIA Novosti, 38 NASA/VRS, 39 Detlev van Ravenswaay, 42 Tommy Eliassen, 43 Pekka Parviainen, 50-51 Mark Garlick, 52 Claus Lunau, 55 John Chumack, 59 Universal History Archive, 62 (B) Detlev van Ravenswaay, 68 David A. Hardy, 69 Christian Darkin, 71 John Sanford, 78 Babak Tafreshi, 79 Babak Tafreshi, 80 ESA, 81 John R. Foster, 87 Sheila Terry, 90-91 Frank Zullo, 93 Monica Schroeder, 95 (T) Monica Schroeder, 95 (B) Chris Cook, 97 (T) Monica Schroeder, 97 (B) SPL, 102 (R) Monica Schroeder, 103 (composite) Frank Zullo, 107 (T) Royal Astronomical Society, 107 (B) Monica Schroeder, 110 (R) Monica Schroeder, 111 (R) Monica Schroeder, 116 Larry Landolfi, 117 Luke Dodd, 122 Dan Schechter, 123 Mark Garlick, 125 Carl Goodman, 126 (T) Manfred Kage, 126 (B) Dennis Kunkel Microscopy, 127 John Chumack, 128 Pekka Parviainen, 130 SPL, 131 John Chumack, 142 Chris Cook, 146 Mark Garlick, 148 Tony & Daphne Hallas, 152 (R) NASA/ESA/XMM-Newton, 153 Mark Garlick, 154 Gerard Lodriguss, 158 (B) John R. Foster, 164 Juan Carlos Casado, 170 (BR) John Chumack, 173 SPL, 174-5 Walter Pacholka, Astropics, 176 John Sanford & David Parker, 177 Gerard Lodriguss, 179 Luke Dodd, 181 John Sanford, 182 (R) Royal Astronomical Society, 183 Larry Landolfi, 184 Mark Garlick, 185 Detlev van Ravenswaay, 203 Royal Observatory, Edinburgh/Anglo-Australian Telescope Board, 217 John Chumack/Science Source; 220 (B) John Sanford, 223 Robert Gendler, 224 Todd Boroson/NOAO/AURA/NSF, 227 Celestial Image Co., 229 (L), NOAO/AURA/NSF, 231 Russell Croman, 275 (T) Herman Heyn, 275 (B) Jean-Charles Cuillandre/CFHT; **Shutterstock:** 11 vchal, 25 spe, 26 Alex Kosev, 28 Sunti, 33 John Panella, 49 Matthew Bechelli, 60 & Back cover (TL) McCarthy's PhotoWorks, 61 Sportlibrary, 114 Therato, 119 (T) sozon, 144 slavo.o, 182 (L) peresanz, 186-7 Viacheslav Lopatin; **Wil Tirion:** 192-3 & Back cover (TR), 212-13, 238-9, 260-61; **U of Arizona:** 250 (B), Sid Leach/Adam Block/Mount Lemmon SkyCenter; **© Michael Vlasov 2017:** 158 (T), 205 (B), 257 (B); **Stanislav Volskiy:** 159; **Frank Zullo:** 65.

Constellation maps by Mapping Specialists, Ltd.